Astronomers' Observing Guides

D0620009

Other Titles in This Series

Peter Grego

Venus and Mercury, and How to Observe Them

Springer

ISBN: 978-0-387-74285-4   e-ISBN: 978-0-387-74286-1

Library of Congress Control Number: 2007937298

Printed on acid-free paper

9 8 7 6 5 4 3 2 1

springer.com

Dedication
For Jacy, my daughter

Acknowledgements

Thanks to Mike Inglis for having asked me to write this book, and for his help and advice as the project got underway. All the staff at Springer in the UK and USA have worked hard to produce this book, and I am deeply grateful to them. My special thanks go to John Watson and Harry Blom. The image contributors to this book have been tremendously helpful and generous – I hope my text does justice to your work, and that others are inspired to observe Mercury and Venus.

About Peter Grego

Peter Grego has been a regular watcher of the night skies since 1976. He observes from his garden in Rednal, UK, with a variety of instruments, ranging from a 127 mm Maksutov to a 300 mm Newtonian, but his favourite instrument is his 200 mm SCT. Grego's primary interests are observing and imaging the Moon and bright planets, but he occasionally likes to 'go deep' during the dark of the Moon.

Grego has directed the Lunar Section of Britain's Society for Popular Astronomy since 1984, and since 2006 has co-ordinated the Lunar Topographical Subsection of the British Astronomical Association. He edits five astronomy publications – *Luna* the journal of the SPA Lunar Section, *The New Moon*, journal of the BAA Lunar Section, the *Newsletter* of the Society for the History of Astronomy, the *SPA News Circulars* and *Popular Astronomy* magazine. In addition, he writes and illustrates the monthly *MoonWatch* column in UK's *Astronomy Now* magazine, is observing advisor and columnist for *Sky at Night* magazine, and maintains his own website at www.lunarobservers.com.

Grego is the author of a number of astronomy books, including *Collision: Earth!* (Cassell, 1998), the *Moon Observer's Guide* (Philips/Firefly, 2004), *The Moon and How to Observe It* (Springer, 2005), *Need to Know? Stargazing* (Collins, 2005), *Need to Know? Universe* (Collins, 2006) and *Solar System Observer's Guide* (Philips/Firefly, 2005). He is a Fellow of the Royal Astronomical Society.

Contents

Section 2 Observing Mercury and Venus

Contents

Contents

Chapter 1

Introduction – A Perspective on Mercury and Venus

Ever since our remote ancestors developed a keen curiosity about the celestial realms above them, human eyes have regarded the planets Mercury and Venus as two bright points of light that flit between the dusk and dawn skies. The word 'planet' is derived from the Greek 'planetes' meaning 'wanderer', and this comprised the five classical planets – Mercury, Venus, Mars, Jupiter and Saturn – plus the Sun and the Moon, making up the 'seven heavenly objects'. Because of their apparently special status in the sky, these seven mysterious wanderers were incorporated into ancient systems of belief and astrology. Needless to say, there has never been any scientific basis for astrology, and no reasoning person could have any basis for thinking that there is any merit in such quaint, though ignorant, cultural peculiarities.

Mercury and Venus were special among the five classical planets, in that their wanderings never took them very far from the Sun. Neither planet can appear opposite the Sun in the sky. From our point of view on the Earth, Mercury always appears relatively near to the Sun – a little more than an outstretched hand's width at most – so close that it can never be seen shining in a truly dark sky. Glistening with a slightly pinkish hue, and never brighter than magnitude –1.3, Mercury is only visible with the unaided eye for a few weeks at a time, alternating between dusk and dawn skies half a dozen times a year. Its rapid motion relative to the celestial background lent it perfectly to an association with the swift god Hermes in ancient Greece, later transformed into the Roman Mercury, the fleet-footed messenger of the gods.

Venus appears to venture around twice as far from the Sun as Mercury, and at its furthest can remain visible in the twilight skies up to 4 hours after sunset (or before sunrise). Always a dazzling pure white object when seen against a dark twilight sky, Venus can appear as brilliant as magnitude –4.4 – bright enough to cast shadows. It is not surprising that this stunning planet took its name from the Roman Venus (equivalent to the Greek Aphrodite), goddess of love and beauty.

It was only after the Copernican notion that the Sun, not the Earth, lay at the centre of the Solar System, that the true reason for the planets' apparent paths though the heavens became clear. Both Mercury and Venus have orbits that lie within that of the Earth, and as a consequence they are known as the 'inferior' planets. Mars, Jupiter and Saturn orbit the Sun beyond the Earth, and they are called the 'superior' planets. The terms 'inferior' and 'superior' are occasionally confused with 'inner' and 'outer' planets. The inner planets are Mercury, Venus,

Earth and Mars – four rocky worlds that orbit the Sun inside the main Asteroid
Belt. Beyond the Asteroid Belt, the gas giants Jupiter, Saturn, Uranus and Neptune
comprise the outer planets.

The fact that Mercury and Venus were physical globes in orbit around the Sun
– substantial worlds in their own right, rather than mere points of light – became
evident soon after the serious application of the telescope to astronomical purposes
in the 17th century. Telescopes reveal that both Mercury and Venus go through a
series of phases, according to the varying angle between them, the Earth and the Sun,
and therefore the amount of sunlit hemisphere that is observable. Venus presents
such a large apparent size that its phases were evident even to Galileo Galilei,
who first noted the phenomenon in December 1610 using his tiny home-made
refracting telescope. Mercury, substantially smaller through the telescope eyepiece
and somewhat more elusive than Venus, required a somewhat larger telescope for
its phases to be clearly discernable; its phases were first discerned by Giovanni Zupi
in 1639.

Further proof that both planets orbited the Sun came with the observation
that they appear to vary in apparent diameter as they go through their phases.
Both planets appear to grow larger as they move away from the Sun following
superior conjunction (the far side of their orbit as seen from the Earth), their phase
diminishing from a near-fully illuminated disc, through half-illuminated phase at
around their greatest apparent distance east of the Sun. Their apparent diameter
continues to grow as they make an ever-narrowing crescent as they approach
inferior conjunction with the Sun. Precisely the reverse sequence occurs as the
planets move west of the Sun, following inferior conjunction.

Despite their 'inferior' tag (which is sometimes mistaken to be some kind of
a planetary quality judgement), the true planetary status of Mercury and Venus
within the Solar System has never been in any doubt. It may be wondered why
distant Pluto, 2,390 km in diameter and attended by three satellites, lost its planetary
title in 2006, while Mercury has retained its title as a fully-fledged planet. According
to the International Astronomical Union (a body that presides over astronomical
definitions and nomenclature, among other things) a planet in our Solar System
is defined as an object in orbit around the Sun which is massive enough for
its own gravity to pull itself into a spherical shape and which, furthermore, has
gravitationally swept up the neighbourhood around its orbit. Mercury fulfils these
criteria, while Pluto fails in that its distant neighbourhood contains numerous
sizeable bodies called Kuiper Belt Objects, several of which are around Pluto's size.

Planet of Mystery

Speculation that there may be a planet orbiting the Sun within Mercury's orbit
– a so-called intra-Mercurial planet – arose among astronomical circles during
the 19th century. The speculation was not without foundation; it was fuelled by
observations of Mercury's orbit, which appeared not behave in precise accordance
with conventional Newtonian gravitational physics. The orbit of Mercury is the most
markedly elliptical of all the planets. Instead of remaining fixed in space, the point
of its perihelion undergoes precession around the Sun. When seen from above, the
planet traces out a complicated 'rosette' pattern in space, like a spirograph sketch,
as the axis of Mercury's orbital ellipse gradually swings around the Sun. Although
perihelion precession is predicted by classical physics, that observed in the motion

of Mercury appears slightly more – 43 arcseconds per century – than could be accounted for. For a time, it looked as though the gravitational influence of a small undiscovered planet, closer to the Sun than Mercury, might be responsible for this inexplicable phenomenon.

There was a good precedent for the speculation. In 1846, the eminent astronomer and mathematician Urbain Leverrier had calculated that the anomalous motions of the planet Uranus were due to the gravitational perturbations of a large unseen planet further out. Leverrier's prediction of the position of this planet led directly to the discovery of Neptune, by Johann Galle and Heinrich d'Arrest at the Berlin Observatory on 23 September 1846.

On 26 March 1859, Edmond Lescarbault, an obscure rural amateur astronomer, reported having observed a dark spot transit the Sun's disc in the space of 1 hour and 17 seconds. Up until that time, the only planetary transits ever observed were those of Mercury (transits taking place 13 or 14 times each century) and Venus (twice per century). After personally visiting Lescarbault and closely examining his extraordinary observational claims, Leverrier was won over. He asserted that the mysterious transiting object had been nothing other than the elusive intra-Mercurial planet and he went on to confidently calculate its orbit around the Sun. Appropriately dubbed 'Vulcan' (after the Roman god of fire and volcanoes), this putative planet was calculated to orbit at an average distance of 21 million kilometres from the Sun – three times closer to the centre of the Solar System than Mercury. If Mercury was a wing-footed messenger, Vulcan was a planetary email – it was calculated to zip around the Sun in a period of just 19 days and 17 hours, in a plane inclined some 12° to the ecliptic.

Proponents of Vulcan claimed that, as seen from the Earth, the planet always remained uncomfortably close to the Sun's glare, being just eight degrees away at its greatest elongations. As such, it could only be detected in transit across the Sun or during times of a total solar eclipse when the brilliant light of the Sun was dimmed for a brief spell, allowing the space around the Sun to be scanned for interlopers. Decades of purposeful searches proved fruitless, and although there were numerous observational claims – some by very respected astronomers – none were able to be scientifically corroborated.

While the predictions made by classical physics appear to hold firm for objects in relatively weak gravitational fields – like the Earth's pull on your body and the motion of the Earth around the Sun – they begin to become unstuck when applied to situations involving strong gravitational fields. Mercury's orbit is a case in point. The small, though nagging discrepancy in Mercury's orbit was finally explained when Einstein's theory of relativity was published in 1915, a theory which deals with the distortion of space-time by massive objects and its effects. Indeed, so closely does the orbit of Mercury fit in with relativistic predictions that there simply couldn't be a substantial intra-Mercurial body, since it would produce a noticeable effect. So ended one of the most intriguing tales in the history of astronomy.

Mercury and Venus Up Close

In terms of its physical appearance, Mercury is the least known of the Solar System's eight major planets. During the three and a half centuries of telescopic observation prior to the advent of the Space Age, the planet yielded only a scant few visual clues as to the nature and appearance of its surface. After half a century of space exploration, during which no fewer than 150 space probes have been dispatched to various parts of the Solar System, Mercury has avoided being investigated up close in a thorough and systematic manner. Only one space probe has ever viewed Mercury in any sort of detail. The United States' Mariner 10 probe made the first of its swift flybys of the planet on 29 March 1974, followed by two more flying visits on 21 September 1974 and on 16 March 1975. Some 2,700 images secured by Mariner 10 showed a total of around half the planet's surface at resolutions of between 100 and 4,000 m. Mercury was shown to be a heavily cratered world with a strong resemblance to our own Moon's impact battered southern highlands, devoid of an appreciable atmosphere.

Being the closest planet to the Sun, Mercury's sunlit face becomes incredibly hot, and it had always been assumed that it was impossible that ice deposits could ever exist on Mercury. But in August 1991 a team of astronomers at the California Institute of Technology obtained radar images of the Mercurian surface using a high power beam sent by the Jet Propulsion Laboratory's Goldstone antenna and received by the Very Large Array radio telescope in New Mexico. As expected, the planet's rough, heavily cratered surface was revealed; in addition, the scientists were surprised to receive strong radar echoes from Mercury's north polar region, which happened to be tilted strongly towards the Earth at the time the observations were made. This clear radar return, imaged as a single brilliant spot around 400 km in diameter, was interpreted as evidence for the existence of a large ice deposit, trapped on the permanently shadowed floors of deep craters near the pole. If it exists, Mercury's polar ice was probably deposited by impacting cometary nuclei. Once vaporized in the impact, the volatiles in the nucleus, including water, would have frozen out in areas shadowed from sunlight.

Despite the dearth of detailed up-close scrutiny, Mercury is a pretty important planet, and planetary scientists are of the unanimous opinion that Mercury has been deserving of far closer scrutiny than it has yet enjoyed. A better understanding of the planet's global topography, its geological processes, composition and internal structure will enable scientists to gain a deeper insight into the development of the Solar System and the processes that took place in the inner Solar System in the remote past. Indeed, it is thought that Mercury's fascinating impact-scarred surface has resulted, to a large extent, from processes worked by the gravitational perturbations of distant asteroids and comets by the Solar System's outer gas giants.

If all goes to plan, the balance will begin to be redressed very soon. NASA's Messenger (MErcury Surface, Space ENvironment, GEochemistry, and Ranging) space probe, launched in August 2004, is to be inserted into a year-long orbit around Mercury, commencing in 2011. On the way, the probe will be given gravity assists from close flybys of Venus (in October 2006 and June 2007) and then Mercury itself in January and October 2008 and September 2009. Scientific observations and images of both planets are to be made during these flybys.

Venus has also deserved the epithet of 'mysterious planet', but for entirely different reasons. When viewed in ordinary light, the planet's solid surface can never be seen from the Earth because it is swathed in a dense, perpetually cloudy atmosphere. The topmost layers of these clouds reflect most of the sunlight falling on them, giving Venus its intense brightness. A succession of telescopic observers, unaware that any features they might have observed were merely transient cloud phenomena, attempted to map the planet. A number of peculiar charts of Venus' imagined surface were produced, some of them extraordinarily detailed. Indeed, some astronomers ventured to determine the rotation period of Venus based on the surface features that they imagined having seen.

Our first real glimpses of Venus' solid surface came in 1964 when radar was used to produce basic images of the planet's surface. Among the first radar surveys was undertaken in 1969 by Richard Goldstein and Howard Rumsey using the Goldstone antenna. The maps, based on studies on 17 dates around the time of Venus' closest approach at inferior conjunction, show around 30 percent of the Venusian surface. The images were coarse, and despite enhancement they appeared somewhat distorted. Similar studies were also made in 1969, using the 305 metre radio telescope at Arecibo in Puerto Rico (a giant antenna built within a natural crater in the hills, which made the first measurement of Mercury's period of rotation in 1965). Brighter, more radar-reflective areas were thought to be hilly or mountainous regions, whilst the darker areas were thought to be smooth plains. The nomenclature initially chosen for Venus was unexciting – two bright spots were called Alpha Regio and Beta Regio. Earth-based radar research continued into the 1970s, gradually becoming more sophisticated to reveal ever finer detail. Circular formations began to be discerned, although it was not possible to determine either a volcanic or impact origin for them. However modest these results, it was a promising beginning.

The US Pioneer-Venus Orbiter began its detailed radar mapping in December 1978, with a resolution far better than was then possible with Earth-based studies. The probe exceeded its planned operational life of a year – it lasted no less than 14 years, eventually burning up as it plummeted through Venus' atmosphere in October 1992. Venus was shown to have two distinct types of terrain. Much of the planet is covered by an undulating landscape of low relief – a monotonous vista accounting for around 80 percent of the surface – from which rise three major highland plateaux.

The US Magellan probe was a natural evolution of Pioneer-Venus, entering orbit in August 1990 to begin its high resolution radar mapping. Features as small as 120 m were discerned, and the height of the surface relief was measured to an accuracy of just 30 m. By September 1992 Magellan had mapped 99 percent of Venus, laying bare a planet which has undergone – and is still experiencing – widespread endogenous activity. Atmospheric and surface information gathered by a host of other probes, including the Soviet Veneras 11, 12, 15 and 16, in addition to a variety of Earth-based studies, leads scientists to believe that major volcanic activity may well be taking place today.

Space probes and soft-landers have shown Venus' surface to be one of the least welcoming in the entire Solar System. Its surface is hot enough to melt lead and the atmospheric pressure is equivalent to being a kilometre under the ocean. In addition there are occasional volcanic eruptions which trigger planet-wide acid rainfall. The first softlander to brave these abysmal conditions was the Soviet Venera 7, which parachuted down in December 1970. Though it was constructed like a tank, it survived for just 23 minutes on the surface before being fried and crushed. Seven more Venera probes softlanded on Venus, some returning pictures of the wild volcanic landscape and sampling the soil. All succumbed to the terrible conditions in a short space of time.

For the present time, and for the foreseeable future, scientific studies of Venus are to be conducted by space probes in orbit around the planet. ESA's Venus Express space probe entered a 24 hour elliptical quasi-polar orbit around Venus in April 2006. During its nominal 500 day mission (around two Venusian sidereal days) the probe will be making a comprehensive series of observations aimed at gaining a greater understanding of how the planet's atmosphere works. A battery of instruments will investigate Venus' cloud systems, atmosphere, magnetosphere, the planet's plasma environment and its interaction with the solar wind. Although the probe carries no radar mapping equipment, some of the planet's surface properties are capable of being measured using the VIRTIS (Visible and Infrared Thermal Imaging Spectrometer) instrument. VIRTIS will search for variations related to the interaction between the surface and atmosphere, map the planet's surface temperature in a search for possible active volcanic hot spots, in addition to searching for seismic waves propagated by acoustic waves amplified in the mesosphere (an atmospheric layer between 65 and 100 km in attitude). One exciting early result from Venus Express was the discovery of large twin atmospheric vortices over the planet's south pole.

Venus' atmosphere is also to be closely investigated by PLANET-C, a climate orbiter to be launched by JAXA (Japan Aeospace Exploration Agency) in 2010. From an elliptical orbit similar to that of Venus Express, the planet's meteorological phenomena will be monitored, as well as infrared imaging of the surface.

Although at the beginning of the 21st century we know a great deal about Mercury and Venus, science has much yet to discover about these two planets nearest the Sun. As we will discover, amateur visual observations, drawings and images of both planets remain valid scientific pursuits, as well as being activities that are thoroughly enjoyable, often challenging in many ways, but always personally instructive.

Peter Grego
Rednal, England, May 2007

Current Knowledge of Mercury and Venus

Chapter 2

Our Current Knowledge of Mercury

Mercury's Orbit

Mercury orbits at an average distance of 57,909,176 km (0.3871 AU, or 3.22 light minutes) from the Sun, in a sidereal orbital period (its period with respect to the stars) of 87.99 days. Its actual distance from the Sun ranges between 46,001,272 km (0.3075 AU) at perihelion and 69,817,079 km (0.4667 AU) at aphelion. Mercury's average orbital velocity is 47.36 km/s. At perihelion it reaches its maximum orbital velocity of 58.98 km/s, while its minimum orbital velocity of 38.86 km/s is attained at aphelion. With an eccentricity of 0.2056, Mercury's orbit is the most elliptical of all the planets, and its inclination by 7.005° to the plane of the ecliptic (3.38° to the Sun's equator) far exceeds that of any other planet.

Viewed from above, Mercury's orbit traces out a complex 'rosette' pattern owing to the advance of the point of its perihelion by 574 arcseconds per century (completing one circuit around the Sun every quarter of a million years, clockwise when viewed from the north). However, this figure differs by some 43 arcseconds per century according to predictions made using Newton's Law of Universal Gravitation. Einstein's General Theory of Relativity explains why Mercury's perihelion point should advance, and allows us to calculate the precise degree of advancement. As Mercury speeds towards perihelion, its velocity increases, and as a consequence its overall relativistic mass increases too. This produces a small acceleration in the planet's orbital velocity, which nudges the perihelion point along very slightly more than that predicted by classical Newtonian physics.

Physical Dimensions

Mercury is a terrestrial planet, composed chiefly of silicate rocks, like all four of the Solar System's inner planets. It is the smallest of the planets and ranks as the 11th largest object in the Solar System.

With an equatorial diameter of 4,879 km, Mercury is a little more than one-third (0.383) the Earth's diameter, and having a volume of 61 billion cubic kilometres it is 1/18th the physical size of the Earth. The precise shape of Mercury is currently unknown because it is too small to measure accurately from the Earth, although it is probably slightly out of spherical, forming a triaxial ellipsoid, with two opposite permanent bulges in the plane of its orbit around the Sun. There are indications that there may be a degree of hemispheric asymmetry, its southern hemisphere being slightly larger than its northern hemisphere. It is inevitable that solar tidal forces produce repetitive crustal deformation, although it is likely to be in the order of around a metre in amplitude due to Mercury's substantial solid crust, which is thought to be 100–200 km thick.

Mercury's surface area of some 75 million square kilometres amounts to some 15 percent that of the Earth, equivalent to the area of the Atlantic Ocean, or twice the area of the Moon. Two planetary satellites – Jupiter's Ganymede and Saturn's Titan, 5,262 km and 5,150 km in diameter, respectively – are actually somewhat larger than Mercury, although both of those distant icy moons are considerably less massive.

The Earth, Moon and Mercury, to scale.

Mass, Density and Gravity

Mercury has no natural or artificial satellites, and therefore its mass cannot be calculated using Kepler's third law. Gravitational perturbations exerted on the Mariner 10 space probe during its three flybys of Mercury has allowed its mass to be determined at 3.30×10^{23} kg (330 trillion tonnes), 1/18th that of the Earth, making it the Solar System's ninth most massive object. Mercury's material has an average density of 5.43 g/cm^3 – second only to the stuff of the Earth, which is slightly more dense, at 5.52 g/cm^3 (0.983 the Earth's density). However, when corrected for the effects of gravity, Mercury is considerably denser than the Earth, having an uncompressed density of 5.3 g/cm^3 compared with 4.4 g/cm^3 for the Earth. Mercury's surface gravity at the equator is 0.284 percent that of the Earth's, and its escape velocity is 4.435 km/s.

Axial Tilt and Rotation Period

Since Mercury's rotational axis is tilted a mere 0.01° to the plane of its orbit around the Sun, the planet experiences no seasons. In Mercury's skies, the north celestial pole lies at 18 h 44 m 2 s RA, 61.45° Declination – in the constellation of Draco, midway between the stars Polaris (Earth's current pole star) and Vega (which, by virtue of precession, is to be the Earth's pole star in around 12,000 years' time). Although there is no bright star marking Mercury's north celestial pole, the star Alpha Pictoris (magnitude 3.31) lies very close to its south celestial pole.

Mercury rotates on its axis just once every 58 days 15 hours 30 minutes (this being termed its sidereal spin period, ie., its rotation with respect to the stars). The rotational velocity of a point at its equator is 10.89 km/hour, around 154 times slower than the rotational velocity at the Earth's equator.

Mercury is locked in a 3/2 spin-orbit resonance, rotating three times on its axis for every two orbits around the Sun. This means that alternate hemispheres face the Sun each perihelion. As a result, the planet has two 'hot poles' – points on the equator, one at 0° and the other at 180° longitude – that lie directly beneath the Sun when Mercury is at perihelion. Temperatures on Mercury's surface range from 740 Kelvin at the hot poles during their exposure to the Sun at alternate perihelia (lead melts at 601 Kelvin) to 90 Kelvin on the planet's dark side. The range of temperatures on the surfaces of planets like the Earth and Venus is moderated by their atmospheres, while Mercury's lack of a substantial atmosphere allows a greater temperature range, its surface rapidly cooling in the night and heating during the daytime.

Following sunrise from a point on the planet's equator, it takes around 44 Earth days (half a Mercurian year) for the Sun to reach its zenith, and a further 44 days until sunset. A complete 'day' on Mercury, from one sunrise to the next, lasts 176 Earth days – three times longer than the planet's rotation period and twice as long as a Mercurian year.

A peculiar kind of sunrise is to be observed from points on the equator at both 90° and 270° in longitude (90° either side of the 'hot poles'), which experience alternating sunrises and sunsets when the planet is at perihelion. From these longitudes on Mercury's surface, the Sun takes around four Earth days to heave most of its disc above the horizon, growing from around 96 to 102 arcminutes in apparent diameter as it does so. At this point, the Sun then falls back towards the horizon for a further eight days, and then reverses it course once more to recommence its rising, now slowly shrinking in apparent diameter; the whole solar disc finally manages to clear the horizon around 18 days after first showing itself. This bizarre solar dance – unique in the Solar System and seemingly counter-intuitive in nature – is caused by the planet's orbit around the Sun near perihelion temporarily canceling-out the observed effects of the planet's rotation. It is noticeable at perihelion from the opposite point on the planet, where sunset is a similarly protracted affair, the sequence of apparent solar motions in Mercury's skies being precisely the opposite as sunrise, while from the 'hot pole' the Sun appears to oscillate around the zenithal point at around perihelion.

Origins

To fully understand the origin of the Sun and its planetary system, it is necessary to go back some 4.6 billion years, when the Universe was two-thirds its current age. Like today, the starry spiral of the Milky Way galaxy was interlaced with giant molecular clouds – vast gravitationally-bound entities, typically around 100 light years across, composed of cold interstellar gas (99 percent – mainly molecular hydrogen, with helium and a small proportion of nitrogen and oxygen) and dust (1 percent – particles of silicates and metals).

In one of these giant molecular clouds, embedded within an outer spiral arm of the Galaxy, a ripple originating from beyond the cloud pushed some of its material together to form a multitude of denser regions. The origin of the ripple which served to trigger the formation of these denser clumps is unknown, but there are a number of possible causes. It may have been produced by the gravitational tug of a massive passing star or star cluster; it may have propagated from a nearby supernova explosion, or it may have been caused by the shockwaves surrounding a growing bubble of hot gas surrounding a highly energetic, but short-lived massive star in the vicinity.

Whatever the cause of the ripple, some of the material of the interstellar cloud was shunted together to form denser regions. Many of these denser regions formed dark globules which collapsed freely in on themselves under their own gravity – perhaps thousands of such entities to arise in the disturbed interstellar cloud. One of these collapsing globules was ultimately to become the Solar System.

Once temperatures at the globule's core had exceeded 10,000 Kelvin, the solar globule became a distinct proto-Sun – an object as wide as Neptune's current orbit, but still far too cool to ignite thermonuclear reactions. Surrounding the rapidly spinning flattened globe of the proto-Sun lay a broad disk of dust and gas lying along the plane of the proto-Sun's rotation, more than 100,000 AU in diameter (1 AU, an Astronomical Unit, is 150 million kilometers, the distance between the Earth and Sun). By around 100,000 years after its formation, the proto-Sun's diameter had shrunk to about the size of Mercury's current orbit.

Near to the nascent Sun, in the region now occupied by the inner planets, temperatures in the protoplanetary disk were too high for lightweight gases such as hydrogen and helium, to condense, but low enough for silicates and heavier metallic elements to condense. Beyond around the current distance of Jupiter, some 5 AU from the nascent Sun, icy material – water, methane and ammonia – condensed out of the disk, forming ice crystals.

Specks of solid matter condensing throughout the disk gently collided with each other, sticking together by means of agglomeration to create chunks of material ranging between one centimetre and ten metres in diameter (something like the same range of particle sizes found in the rings of Saturn today). Material in the protoplanetary disk encircling the nascent Sun began its gravitational accretion into innumerable larger clumps of matter – the seeds of planetary formation – during this period. Gravitational forces enabled the larger objects to snowball in size, enabling mountain-sized planetesimals to form. At this stage, the pace of the growth of the larger planetesimals quickened, producing protoplanetary objects that became the nuclei of the planets.

In the outer Solar System, where the initial process of agglomeration was helped by the fact that ice particles bond together much more efficiently than silicates, the ice-rock cores of four very large protoplanets developed. Proto-Jupiter was first to reach a mass of around a dozen Earths – massive enough to gravitationally sweep up the gases (mainly hydrogen and helium) in its orbital vicinity until it attained a mass of some 30 Earths. Following in a similar fashion, proto-Saturn, proto-Uranus and proto-Neptune grew more massive as their ice-rock cores surrounded themselves with thick gaseous atmospheres.

By around one million years after the collapse of the solar globule, the proto-Sun crossed a significant threshold to become what is known as a T-Tauri star. Although its core temperature of around 5 million degrees was not sufficient to ignite thermonuclear reactions, its energy output, fuelled by gravitational collapse, was prodigious but variable, prone to sudden flares as material was sucked into the star along powerful magnetic field lines. Strong solar winds slammed into its surrounding protoplanetary disk and were deflected perpendicular to its plane, producing a bipolar flow that was eventually to channel a large proportion of the star's original mass into interstellar space. The further growth of the Solar System's giant planets was stopped in its tracks, as any remaining gas was swept away by the solar wind.

Still contracting under gravity, pressures and temperatures at the nascent Sun's core became so high – in the order of 15 million Kelvin – that thermonuclear reactions were triggered, and the Sun was born. It is thought that the entire process, from the beginnings of the collapse of the solar nebula to the time the Sun ignited as a star, took less than 50 million years.

Countless smaller planetesimals remained at large in the Solar System – many of these survive to this day as asteroids and comets. Planetesimals orbiting the Sun between Mars and Jupiter were unable to form a single planet-sized object because of the gravitational disruption of Jupiter, and the objects remained as individual asteroids within the main Asteroid Belt (its total mass being somewhat less than our own Moon).

Remnants of the Sun's protoplanetary disk comprise the Kuiper Belt, which occupies a zone spreading beyond Neptune to around 50 AU and lying roughly along the ecliptic plane. However, vast numbers of icy planetesimals were flung far beyond the Solar System after close encounters with Jupiter and Saturn, to populate a spherical zone between 50,000 and 100,000 AU from the Sun, known as the Oort Cloud. The Oort Cloud is estimated to contain billions of icy comets and has a total mass of between five and a hundred Earths.

By around 200 million years after the collapse of the solar globule, four sizeable protoplanets dominated the inner Solar System – Mercury, Venus, Earth and Mars – each one composed chiefly of silicates and metals. None of the terrestrial planets was large enough to have attracted a disk of material substantial enough to form its own satellite, as the giant outer planets had done. Phobos and Deimos, Mars' two satellites, are tiny captured asteroids – objects like them are likely to have temporarily orbited Mercury, Venus and the Earth before being tidally disrupted and crashing down to their surfaces.

Core, Mantle and Crust

As we have seen, Mercury formed from a limited selection of raw materials in a zone close to the hot young Sun. After formation, the planet was a single homogenous mass, but heating enabled the material to melt and separate according to its density, a process known as differentiation. There were several sources of this heating: internal pressure, radioactive decay of elements and heat produced by asteroidal impacts. Differentiation caused heavier iron-rich material to sink towards the planet's centre, forming the planet's core, while lighter material such as silicon, magnesium and aluminium rose to form the planet's mantle and crust. All the terrestrial planets have a similar internal structure, namely a metallic (mainly iron) core, surrounded by a silicate mantle and overlain by a solid rocky crust.

Mercury's high mean density results from its composition, thought to amount to some 70 percent iron and 30 percent silicates. Most of the iron is contained within an enormous core measuring up to 3,900 km across (around half the planet's volume, or up to 80 percent its diameter) while silicates make up much of the mantle and crust.

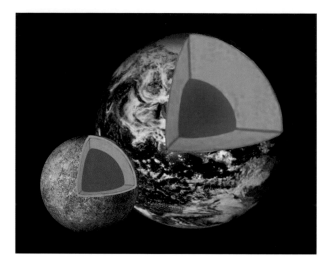

Comparison between the interiors of the Earth and Mercury.

Following differentiation, it is entirely possible that Mercury's iron core was further augmented by the impact of another planetary body whose own iron-rich core was subsumed into that of Mercury, evaporating any remaining volatiles in the process, while vast quantities of the lighter crust and mantle of both objects were flung into space (to be largely consumed by the Sun). Simulations featuring a glancing impact by a body one-sixth the mass of proto-Mercury at a velocity of 126,000 km/hour show that most of the two bodies' rocky material is lost to space while the iron-rich cores melded into a single mass, ultimately producing a body of similar mass and composition to today's Mercury.

A Tenuous Atmosphere

Being such a small planet so near to the Sun, it is unsurprising that Mercury's atmosphere is far less substantial than that surrounding Venus, the Earth or Mars. Any substantial atmosphere it may once have had after its formation soon dissipated into space.

With a surface pressure of around 10^{-15} bars (a trillionth the Earth's atmospheric pressure at sea level), Mercury's atmosphere is almost indistinguishable from a vacuum. In terms of composition, it is made up of oxygen (42 percent), sodium (29 percent), hydrogen (22 percent), helium (6 percent), calcium (0.5 percent) and potassium (0.5 percent), although these are not firmly constrained values. Traces of carbon dioxide, water, nitrogen, argon, xenon, krypton and neon are also likely to exist.

However, Mercury's atmosphere is by no means a stable envelope of gases – atoms are continually being lost and replenished. Atoms of hydrogen and helium are probably derived from the solar wind, gathered by the planet's magnetosphere for a while before later drifting back into space. Other atmospheric constituents are liberated from the planet's crust by the impact of energetic photons and ions from the Sun and by micrometeoritic impacts. Helium, sodium and potassium are also provided by the radioactive decay of elements within Mercury's crust, while degassing events through deep-seated fissures in the crust may occasionally belch out quantities of sulphurous gases. Occasional impacting cometary nuclei undoubtedly liberate quantities of water vapour and other volatiles which freeze out on the planet's unilluminated hemisphere – most of this sublimates as soon as the surface is subjected to sunlight, but there are strong indications that the floors of deep, permanently shadow-filled craters at the north pole are permanently covered with water ice.

Magnetic Field

Mercury has a substantial dipole magnetic field, with positive and negatively charged poles, like the Earth. Also in common with the Earth, Mercury's magnetic axis is roughly in line with the planet's axis of rotation; in Mercury's case the axes are within 14° of each other. The planet's magnetic field strength amounts to some 0.002 Gauss – around 1/100th the strength of the Earth's magnetic field. This implies that it is being generated by means of an internal dynamo mechanism – mechanical energy being converted to magnetic energy – but how this might be produced within Mercury is poorly understood. If Mercury is the smallest and most slowly rotating planet to possess an active magnetic dynamo, most current models for its existence require a molten core. While the core of Mercury is thought to have solidified long ago, it is possible that the planet's solid crust is separated from the inner core by a fluid mantle, and that it has an outer liquid core, giving rise to a dynamo effect. Alternatively, it has been suggested that Mercury's magnetic field emanates from remnant magnetisation of a crust rich in iron, or that it is somehow induced by interactions between the Sun's magnetic field and that of the planet. Both of these models, especially that of intrinsic fossilised Mercurian magnetism, fail to account for the observed magnitude of the magnetic field strength.

While Mercury's magnetic field is far too weak to produce a belt of trapped energetic charged particles (such as the Earth's Van Allen Belts), the pressure of the solar wind distorts the geometry of the planet's magnetosphere. Mercury's magnetic field extends some distance out into space and deflects the solar wind, slowing it down and creating a shock wave known as a bow shock. The bow shock's nose (the portion in line with and facing the Sun) is around 1.5 Mercury radii distant from the planet's surface, with the nose of the magnetopause (the boundary of the planet's magnetic field) a little less than one Mercury radius away from the surface. On the leeward side the magnetic field is stretched out into a long magnetic tail perhaps 15 Mercury radii in length.

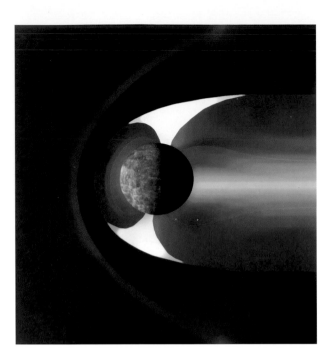

Mercury's intrinsic dipole magnetosphere.

Surface History of Mercury

Mercury's geological history has been arranged into several distinct periods based on relative ages, namely: the Pre-Tolstojan (PT), Tolstojan (T), Calorian (C), Mansurian (M) and Kuiperian (K) Periods. Tolstoj and Caloris are large multi-ring asteroidal impact basins whose main rings are 510 km and 1,340 km across respectively; Mansur and Kuiper are impact craters measuring 100 km and 62 km in diameter respectively.

The Pre-Tolstojan period begins with the earliest phases of the planet's history around 4.6 billion years ago. Global melting of the outer several hundred kilometres of Mercury's original crust may have produced a 'magma ocean' similar to that thought to have covered our own Moon. Such an episode would have produced further differentiation, allowing low-density plagioclase minerals to rise to the highest levels in the crust, forming a predominantly anorthositic crust like that of the Moon.

Conditions on the early Mercurian crust were hardly stable; meteoroid and asteroid impacts continued to smash into the planet, 'sandblasting' the original Mercurian crust over time and producing highly cratered terrain, which makes up the planet's oldest geological units. This ancient highly cratered terrain is still evident across tracts of the current surface of Mercury. All but two of the 23 currently known large Mercurian multi-ring basins (listed below) were formed by asteroidal impacts during the Pre-Tolstojan Period.

These ancient Mercurian impact basins and their concentric rings have been identified using a variety of criteria, including the identification of mountain ranges and isolated massifs that protrude through geologically younger units, arcuate lobate ridges and scarps and isolated topographic features of high relief found in heavily cratered areas. Similar criteria have been used in the identification

Basin	Centre	Age	Ring diameters (km)
1. Caloris	30N, 195W	C	630, 900, **1340**, 2050, 2700, 3700
2. Tolstoj	16S, 164W	T	260, 330, **510**, 720
3. Van Eyck	44N, 159W	PT	150, **285**, 450, 520
4. Shakespeare	49N, 151W	PT	200, **420**, 680
5. Sobkou	34N, 132W	PT	490, **850**, 1420
6. Brahms-Zola	59N, 172W	PT	340, **620**, 840, 1080
7. Hiroshige-Mahler	16S, 23W	PT	150, **355**, 700
8. Mena-Theophanes	1S, 129W	PT	260, 475, **770**, 1200
9. Tir	6N, 168W	PT	380, 660, 950, **1250**
10. Andal-Coleridge	43S, 49W	PT	420, 700, 1030, **1300**, 1750
11. Matisse-Repin	24S, 75W	PT	410, **850**, 1250, 1550, 1990
12. Vincente-Barma	52S, 162W	PT	360, **725**, 950, 1250, 1700
13. Eitoku-Milton	23S, 171W	PT	280, 590, 850, **1180**
14. Borealis	73N, 53W	PT	860, **1530**, 2230
15. Derzhavin-Sor Juana	51N, 27W	PT	**560**, 740, 890
16. Budh	17N, 151W	PT	580, **850**, 1140
17. Ibsen-Petrarch	31S, 30W	PT	425, **640**, 930, 1175
18. Hawthorne-Riemenschneider	56S, 105W	PT	270, **500**, 780, 1050
19. Gluck-Holbein	35N, 19W	PT	240, **500**, 950
20. Chong-Gauguin	57N, 106W	PT	220, 350, 580, **940**
21. Donne-Molière	4N, 10W	PT	375, 700, 825, **1060**, 1500
22. Bartók-Ives	33S, 115W	PT	480, 790, **1175**, 1500
23. Sadi-Scopas	83S, 44W	PT	360, 600, **930**, 1310

Relative ages: C – Calorian Period; T – Tolstojan Period; PT – Pre-Tolstojan Period. Diameter in **bold** is the basin's main topographic rim. Data derived from *The Geology of Multi-Ring Impact Basins*, Paul Spudis, Cambridge University Press, 1993.

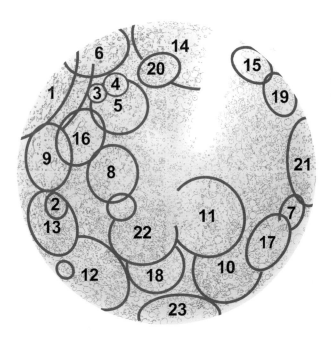

Hemispherical map (Centred on longitude 10W) showing the main rims of Mercury's known major impact basins.

of ancient, eroded multi-ring basins on the Moon. During this period of heavy asteroidal bombardment, molten material burst through weak points in the crust to spread across the surface as lava flows, extensively infilling the existing basins and producing some of the broad intercrater plains which covered or masked much of the pre-existing cratered landscape. The intercrater plains are the most widespread form of Mercurian landscape. This widespread episode of volcanism was at its height around four billion years ago.

Events further out in the Solar System appear to have had enormous repercussions on the terrestrial planets, including Mercury. Millions of asteroids are thought to have been gravitationally disrupted by the Solar System's gas giants – perhaps triggered by a change in the orbit of Jupiter, disrupting the Asteroid Belt, or the late formation of Uranus and Neptune which disrupted a zone of debris further afield. Millions of asteroids were diverted towards the inner Solar System, producing in an intense period of impacts affecting Mercury, Venus, the Earth and Moon that lasted from around 4 to 3.8 billion years ago, and known as the Late Heavy Bombardment.

The Late Heavy Bombardment saw the beginning of the Tolstojan Period, with the formation of the Tolstoj multi-ring basin near the equator, and ends around 3.8 billion years ago with the Caloris impact, one of the most geologically significant markers in Mercury's history. The Caloris impact produced a vast basin whose concentric rings spread across much of an entire Mercurian hemisphere. Numerous lineaments, valleys and crater chains radial to Caloris have also been identified. The most obvious topographical element of the Caloris basin today is the feature's main ring, some 1,340 km in diameter, marked by the Caloris Montes. Immense compressional seismic waves produced by the Caloris impact traveled through the crust and around the globe, in addition to traveling directly through the planet's mantle and core, focusing their energy on a point antipodal to the centre of Caloris. These concentrated shock waves disrupted and destroyed pre-existing topographical features, producing a 'weird' landscape of knobbly hills.

A different form of activity, deep inside Mercury, had further implications on the planet's surface topography. As the core cooled it shrunk, and as a consequence the diameter of Mercury reduced by around three kilometres. Enormous global crustal stresses caused the extensive formation of thrust faults as the crust crumpled up on itself. These faults cut cleanly through pre-existing topography, mountains and craters alike, leading to huge cliffs and lobate scarps, typically several hundred kilometres long and up to three kilometres high.

Mercury's youngest terrain consists of smooth plains, and these occupy around 15 percent of the planet's surface. Low in crater numbers and crossed here and there by compressional features like wrinkle ridges, the smooth plains have a uniform composition. It has been suggested that the smooth plains are thick blankets of impact ejecta, similar to the Cayley Formation found on the Moon. However, evidence for their mode of origin strongly favours a global episode of volcanism triggered by the Caloris impact, when smooth volcanic plains, such as the Suisei, Odin and Tir Planitiae, spread across parts of the planet's surface.

Following this final episode of volcanic activity, the planet's crust consolidated yet further and its interior cooled down and largely solidified. Mercury has remained geologically quiet, apart from occasional meteoroid, small asteroid and comet impacts. An absence of large-scale volcanic activity and rapidly diminishing rates of impact marks the Mansurian Period, which lasted from around 3.5 to 1 billion years ago. During this period it is likely that much of the ice currently existing within permanently shadow-filled craters near the planet's north pole was delivered by the impacts of cometary nuclei. The impact formation of the crater Kuiper around one billion years ago marks the start of the current era. Like many of the relatively small young craters that formed during this period, Kuiper displays an extensive system of bright radial spidery rays.

Some notable examples of Mercury's geological features

Type	Example	Size (km)	Location	Type	Age
1. Impact crater (Kuiperian)	Kuiper	d.62	11S, 31W	I	K
2. Impact crater (Mansurian)	Mansur	d.100	48N, 163W	I	M
3. Impact crater (Calorian)	March	d.70	31N, 176W	I	C
4. Impact crater (Tolstojan)	Ibsen	d.159	24S, 36W	I	T
5. Lobate scarp (rupes)	Discovery Rupes	l.550	56S, 38W	T	C
6. Valley (vallis)	Goldstone Vallis	l.150	16S, 32W	I	PT
7. Smooth plain	Dostoevskij	d.411	45S, 176W	I/V	PT/C
8. Dark halo crater (DHC)	Basho	d.80	33S, 170W	I	K
9. Mountains (montes)	Caloris Montes	l.2,000	39N, 187W	I/T	C*
10. Hummocky plains	Caloris Planitia	w.800	31N, 190W	V/T	C
11. Lineated terrain	Area around van Eyck	w.800	43N, 159W	I	C
12. Weird terrain	Area near Petrarch	w.800	30S, 30W	T	C
13. Ridge (dorsum)	Schiaparelli Dorsum	l.300	23N, 164W	T	C
14. Ancient multi-ring basin	Budh	d.850	17N, 151W	I	PT
15. Multi-ring basin	Caloris	d.1,340	30N, 195W	I	C

Type: I – impact; V – volcanic; T – tectonic.Relative ages: K – Kuiperian Period; M – Mansurian Period; C – Calorian Period; T – Tolstojan Period; PT – Pre-Tolstojan Period.

*The length of the known eastern sections of the Caloris Montes. The actual extent of the mountain range is currently unknown, although it is estimated to have a total length of around 4,000 km if it circumscribes the entire Caloris basin.

Big Mercurian craters (between 10W and 190W longitude, larger than 200 km)

Name	Diameter (km)
Beethoven	643
Dostoevskij	411
Tolstoj	390
Goethe	383
Shakespeare	370
Raphael	343
Homer	314
Monet	303
Vyasa	290
Van Eyck	282
Mozart	270
Haydn	270

(Continued)

Name	Diameter (km)
Renoir	246
Pushkin	231
Rodin	229
Valmiki	221
Wren	221
Michelangelo	216
Bach	214
Mendes Pinto	214
Vivaldi	213
Sholem Aleichem	200

Some bright Mercurian ray craters

Name	Notes
Kuiper	62 km diameter.
Unnamed	Located at 30S, 49W.
Snorri	19 km diameter.
Copley	30 km diameter.
Unnamed	20 km crater west of Chekhov.
Brontë/Degas	Bright pair, 63 km/60 km diameter respectively.
Small unnamed crater Located at 8S, 105W, rays radiate in all directions for several hundred km.	20 km crater.
Unnamed	Small brilliant double crater surrounded by a bright patch (30S, 49W).

Mercurian Nomenclature

Like all the terrestrial planets and their satellites, and the satellites of the gas giants and the asteroids that have been imaged up close, Mercury's surface displays a range of permanent topographic features. With the exception of our own Moon, whose topographic features (at least, those on the side facing the Earth) are readily discernable from the Earth with the play of sunlight across the lunar disk, our knowledge of these topographic features has only come about during the last half a century, when they were imaged at close quarters by interplanetary space probes. Before then, our knowledge of the appearance of the surfaces of the more distant objects in the Solar System depended on visual telescopic observation and photography.

Yet the keenest eyes at the telescope eyepiece and the best Earth-based images could only discern areas of different brightness on the planets (and, in some cases, on Jupiter's four Galilean moons). These variations in tone were caused by areas of different albedo (the actual reflectivity of an area of the surface), plus the angle at which the Sun illuminated them. The latter is especially relevant in the case of Mercury and Venus, inferior planets that display a marked phase and whose terminator (the division between the illuminated and unilluminated parts of the planet) often shows a pronounced darkening.

It may be noted that in some exceptional instances, some of the actual topographic features of Mercury and Mars have been glimpsed visually from the Earth. With regard to Mercury, some observers reported discerning an irregular terminator and even shadow-filled craters, such as the Moon's southern highlands might appear through a very small telescope or opera glasses. These observations, however reliable the observer, were so rare, inconsistent and lacking in verifiable data that they were never incorporated into the 'official' maps of the planet. Therefore, all the pre-Mariner maps of Mercury made using visual observations were based around the planet's larger and more frequently observed albedo features.

A number of notable observers produced their own Mercurian charts upon which they bestowed their own systems of nomenclature. Maps produced by Giovanni Schiaparelli (1890) and Eugene Antoniadi (1934), along with much of the nomenclature devised by the latter, served as the basis for later charts of the planet which were sanctioned by the IAU (International Astronomical Union) prior to the historic Mariner 10 flybys. Notable among these were the maps by Henri Camichel and Audouin Dollfus (1955) and Dollfus and John Murray (1971).

Antoniadi's nomenclature was based upon the mythology of ancient Egypt and Greece. For example, one large dusky area south of the equator was given the grand sounding name of Solitudo Hermae Trismegisti (Wilderness of Hermes the Thrice Greatest), while other dark patches were variously named Solutudino Perse-phones, Solitudino Atlantis and Solitudino Aphdrodites, among others. Brighter spots between these areas were given names such as Pleias, Liguria, Phaethontias, Pentas and Apollonia.

Sadly, these albedo markings were found to bear little relation to the topography of Mercury – quite unlike the surface of the Moon where the dusky markings correspond to lava-flooded multi-ring impact basins, and most of the bright areas are cratered highlands. On Mercury, the ancient flooded basins, smooth terrain and intercrater plains are not that widely different in terms of albedo to the more heavily cratered and mountainous terrain. Rather, the juxtaposition of the observed markings on Mercury is largely due to the overlying patchwork of the planet's bright ray systems which emanate from young impact craters.

At their 1973 Annual General assembly the IAU wisely abandoned the old system of nomenclature and put in its place one devised by the Working Group for Planetary System Nomenclature. This is the system of nomenclature that is referred to in the following survey of Mercury's topography. However, the astronomical world still recognizes the importance of amateur ground-based telescopic work, and astronomical organisations sanction the use of an albedo-marking based nomenclature which is based upon that of Antoniadi. This is similar to the situation with regard to Mars, a planet whose many albedo markings often bear no relation to the planet's topographic features; yet telescopic observers of the red planet prefer to use the old nomenclature devised by Schiaparelli, while planetary scientists use the modern IAU-approved system based upon the planet's topography.

Most craters on Mercury take their names from famous authors, artists and musicians. For example, we find none other than the great playwright William Shakespeare and the artists Jan Van Eyck and Ustad Mansur huddled together in the northern hemisphere, while Ludwig van Beethoven is very nicely composed in the southern hemisphere, his nearby audience including Mark Twain and Arnold Schoenberg. Two notable exceptions to the rule of creative crater personalities on Mercury are craters named Kuiper and Hun Kal. Gerard Kuiper of the University of Arizona was a member of the original Mariner 10 Imaging Team who died shortly before the probe's first flyby of Mercury. Hun Kal is the ancient Mayan name for the number 20 (the Mayas based their numbering system around 20), and was chosen as the crater through which the 20° line of longitude passes; the crater's centre is used to define the system of Mercurian longitudes, in the same way that Greenwich is used on the Earth.

Currently the IAU recognizes generic terminology for 40 different types of feature. Out of this list, five have been applied to Mercurian features, namely Vallis, Rupes, Dorsum, Planitia and Montes.

Vallis (plural Valles), a valley. Mercurian valleys are all named after radio telescopes on the Earth. They include Arecibo Vallis, Goldstone Vallis and Haystack Vallis, each of which figures prominently in the radar mapping of the planet's surface.

Rupes (plural Rupee), a scarp. Scarps are named after famous vessels used in exploration and scientific research, and they include Discovery Rupes (after Captain Cook's final ship that took him across the Pacific Ocean), Victoria Rupes (after Magellan's ship) and Vostok Rupes (after Bellinghausen's Antarctic ship).

Dorsum (plural Dorsa), a ridge. Ridges do not take their names from any specific group, but two are named Antoniadi Dorsum and Schiaparelli Dorsum after the aforementioned astronomers, both of whom played a significant role in the mapping of Mercury.

Planitia (plural Planitiae), a plain. Names for the planet Mercury in a variety of languages, along with Mercury-associated gods from ancient cultures have been adopted for naming the plains. They include Odin and Tir Planitiae (a Norse Mercury god and the Norse word for Mercury) and Budh Planitia (Hindu word for Mercury). There are exceptions; Borealis Planitia (Latin for Northern Plain) is derived from a classical name for a prominent albedo feature, and Caloris Planitia (Latin for Hot Plain) is a reference to its position at one of the planet's hot poles.

Montes (singular Mons), a mountain range. Only one Mercurian mountain range, Caloris Montes, named after the plain that it borders, has thus far been officially designated. However, it is likely that dozens more mountain ranges, in addition to single mountain massifs will be designated by the IAU in the future.

A Survey of the Known Mercurian Surface

Mercury has been described as a paradoxical planet, resembling the Moon on the outside, but like the Earth on the inside. While much of Mercury's surface superficially resembles that of our own Moon, there are significant differences in the range of topographic forms found on Mercury and their modes of origin.

On casual glance, the first thing that is noticed is the heavily cratered nature of the planet's surface, the majority of craters displaying all the hallmarks of being formed by asteroidal impact. Many of the craters are obviously very ancient, highly eroded and overlain with younger features, including younger craters and their ejecta; the floors of many of these ancient craters have also been covered with lava flows. Large multi-ring impact basins in their raw form or filled with lava like those found on the Moon may not be immediately obvious at first sight. Closer examination of Mercury's surface reveals a complex history of bombardment – some of it on a huge scale – plus two distinct episodes of lava flooding, combined with planet-wide crustal adjustments brought about by the planet's shrinkage.

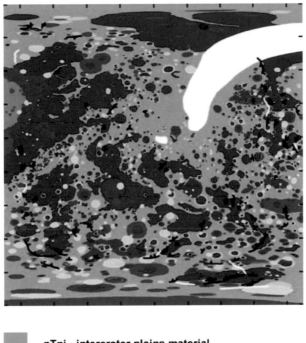

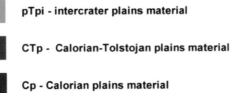

pTpi - intercrater plains material

CTp - Calorian-Tolstojan plains material

Cp - Calorian plains material

Geological chart of Mercury's known hemisphere, showing the most conspicuous units in a complex history of early crustal consolidation, impacts (including numerous major basin-forming events), lava flows, crustal deformation and faulting.

Some of the younger craters lie at the centre of prominent light coloured ray systems comprising ejecta thrown out by the impact process. These ejecta systems are made up largely of pulverised bedrock, plus some traces of the original impactor which was vaporised on impact – this material having itself impacted on its surroundings, forming secondary impact features and churning up the Mercurian regolith. The spindly nature of the rays and their small extent compared to young lunar ray craters of a similar size attests to the effects of Mercury's gravity, which is twice that of the Moon.

Around half of Mercury's surface features have been imaged and mapped with certainty, and the following survey of the planet's topographic features is constrained to these areas. Speculation is certainly possible, but it is perhaps wisest to await further close-up images (which will hopefully be secured some time in the near future) before the mysteries of Mercury's hidden half are answered with any authority.

A Mercurian Quartet

For this detailed topographic survey of the planet Mercury, the planet's known hemisphere (that which was imaged in detail by Mariner 10) has been divided into four regions of approximately equal area, each covering around 10 million square kilometres in extent.

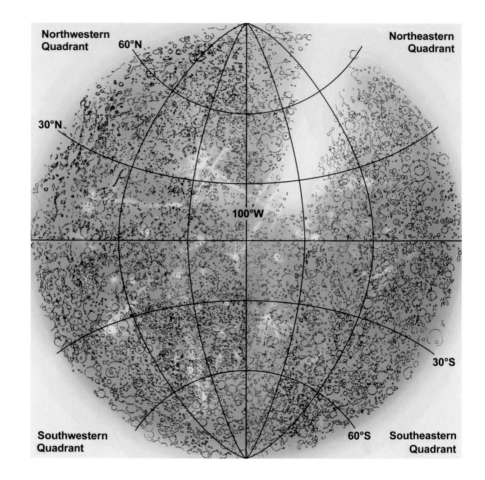

We commence our survey with an examination of the Stravinsky Region (Northeastern Quadrant), a vast swathe of the planet's northern hemisphere from the equator to the north pole, from 10W to 100W. This is followed by an examination the Renoir Region (Southeastern Quadrant) to its immediate south, covering an area from the equator to the south pole, from 10W to 100W longitude. Next comes the Shakespeare Region (Northwestern Quadrant) in the planet's northern hemisphere from the equator to the north pole, from 100W to 190W longitude. Finally, to its south, we describe the Beethoven Region (Southwestern Quadrant), south of the equator and ranging from 100W to 190W.

Each of these four regions is surveyed in a general east to west, north to south trend, using the larger topographic features as the main points of reference in the body of the text; other features near to these main references are surveyed in a general clockwise trend from north, through east. To produce a flowing narrative, these rules are necessarily of a general nature only, and there are a number of diversions and some overlaps with adjoining regions where necessary. Initial references to each particular feature are in **bold face** and in most cases are followed by the latitude and longitude (in parentheses) of the feature's central point, to the nearest degree; this information is often accompanied by the feature's diameter.

Feature names, named feature coordinates and the dimensions of named features have been derived from the US Geological Survey's Astrogeology Research Program website Gazetteer of Planetary Nomenclature at http://planetarynames.wr. usgs.gov/index.html

Accompanying each quadrant line map is a tour map showing the general course of our narrative survey of Mercury's topographical features. Providing the text is read with occasional reference to these maps and the images that accompany it, the reader will find themselves in little danger of getting lost on this enigmatic, crater-crowded little planet.

Northeastern Quadrant

The Stravinsky Region: Mercury, the Planetary Firebird

The Northeastern Quadrant covers a large portion of the planet's northern hemisphere, from the equator to the north pole, from longitude 10W to 100W. Around 1/10th of this quadrant – a tract with an area of around a million square kilometres, around 500 km broad and running northeastward from the crater **Vivaldi** (15N, 86W/213 km) to latitude 70N, longitude 10W – is presently unknown due to deficiencies in the photographic coverage. Of the known face of Mercury, the Northeastern Quadrant is the least cut through by multiple rings concentric to ancient major impact basins. A broad, continuous area north of the equator and covering around half of this quadrant has no traceable basin rings cutting through the crust.

The vast expanse of **Borealis Planitia** (73N, 80W) dominates the entire northern section of this region. With an area of around 750,000 square kilometres, this smooth plain may have been formed (along with a number of Mercury's other planitiae) by a late episode of global volcanism induced by the Caloris impact, around 3.8 billion years ago. The highly cratered nature of the pre-existing topography is evident in the large number of dorsa that appear to trace the shapes of crater rims buried beneath the Borealis lava flows. Bright steaks cross Borealis from a number of directions, and the entire plain is peppered with numerous young impact craters that show up as tiny bright spots, of similar appearance to the lunar crater Linné and its collar of bright ejecta in Mare Serenitatis.

Embedded within northern Borealis, **Goethe** (79N, 45W) is an ancient crater some 383 km across. Its wall is low and considerably eroded, except for the south where it has been breached and submerged by lava flows that extend from the south to cover the crater's floor. The northern and western sections of the wall are wide and numerous structures indicative of terracing, with some surrounding radial lineaments and chain craters on its flanks. Goethe's floor is fairly smooth and flooded, its lava infill probably arising from a later phase of volcanism. As well as numerous small wrinkle ridges, some of which appear to outline buried craters,

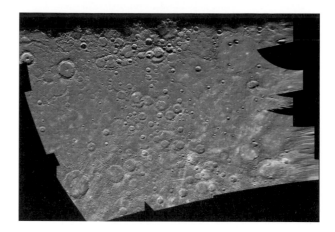

The Borealis region consists of an ancient cratered terrain smothered with extensive lava plains, peppered with more recent impacts. NASA.

several fresher craters and punctuate the floor. Streaks of ejecta cross Goethe; they may originate from a bright young (unnamed) 15 km diameter impact crater which lies some 70 km southwest of **Gauguin** (66N, 96W/72 km), some 700 km away southwest of Goethe. Just beyond Goethe's northern rim lies **Aristoxenus** (82N, 11W/69 km), which currently has the distinction of being the most northerly named crater on the planet.

South of Goethe lies the deep, prominent **Tung Yüan** (74N, 55W), an unusual elongated crater 64 km long with blocky external ramparts and central peaks; its shape may have resulted from the superposition of one large impact crater upon another, or perhaps the near-simultaneous impact of a double asteroid. Beyond Goethe's western rim is **Despréz** (81N, 91W) a well-preserved impact crater some 50 km across, similar in appearance to the lunar crater Bullialdus.

In the southern part of Borealis lies **Monteverdi** (64N, 77W), a 138 km diameter crater whose northern wall has been obscured and overlain by Borealis lava flows; its southern rim, however, remains sharply-defined. To its south is **Rubens** (60N, 74W), an ancient crater some 175 km across whose highly cratered floor has escaped the later lava flooding that affected its northern neighbour.

Immediately to the east of the *terra incognita* the terrain consists of the smooth plains of Borealis but it becomes increasingly cratered; the crater **Sor Juana** (49N, 24W/93 km) marks part of Borealis' irregular eastern border. The density of small craters in this region is at its greatest surrounding **Monet** (44N, 10W), a 303 km flooded basin which lies at the eastern limit of our current knowledge. The area is riddled with ridges and faults that indicate the presence of two ancient basins. The **Derzhavin-Sor Juana** basin (unofficial name) is centred at 51N, 27W and has a main rim some 560 km across. The centre of the other basin, **Gluck-Holbein** (unofficial name) has a main rim of around 500 km across and is located at 35N, 19W. Among the smaller craters in this area are **Gluck** (37N, 18W/105 km), **Echegaray** (43N, 19W/75 km) and **Grieg** (51N, 14W/65 km). Many of these craters are well-defined, with sharp rims and central elevations.

On the eastern side of the aforementioned tract of Mercurian *terra incognita*, Borealis Planitia is cut through by **Victoria Rupes** (51N, 31W) a huge scarp that stretches 400 km south to skirt the eastern external ramparts of the 159 km crater **Derzhavin** (45N, 35W). This huge, somewhat sinuous fault feature appears to be geologically linked with **Endeavour Rupes** (38N, 31W) to its immediate south. Endeavour Rupes curves southward for a further 300 km, terminating a short

The Goethe-Desprez-Tung Yuan region. NASA.

distance west of the crater **Holbein** (36N, 29W). Both Derzhavin and Holbein have substantial walls that display internal terracing and external ridges and furrows, and both have relatively young lava-filled floors. Further south, **Antoniadi Dorsum** (25N, 31W) snakes its way through cratered terrain, cutting through a number of substantial topographic features. This is evidently a compressional ridge formed when Mercury's diameter shrunk slightly, crumpling the crust and distorting older topographic features in the process.

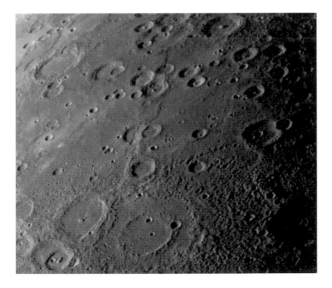

Antoniadi Dorsum. NASA.

A highly eroded double-ringed crater, **Wren** (24N, 35W/221 km) – one of the least distinct of all named Mercurian features – lies to the west of Antoniadi Dorsum. To its northwest and west is an indistinctly charted area which is overlain with long, bright steaks of ejecta oriented southwest-northeast. Here can be found a number of large craters, including **Hugo** (39N, 47W/198 km), **Velázquez** (38N, 54W/129 km), **Kuan Han-Ch'ing** (29N 52W/151 km) and **Praxiteles** (27N, 59W/182 km). The latter is a particularly bright feature owing to its dusting of bright material.

A relatively smooth unnamed plain covers a large area southeast of Holbein. To the south of this plain is the impressive double-ringed basin **Rodin** (21N, 18W); its outer rim measures 229 km across, and its inner ring is 116 km in diameter. Both rings have been flooded with lava, but the inner ring appears smoother and it may have experienced a further phase of flooding. There is extensive external

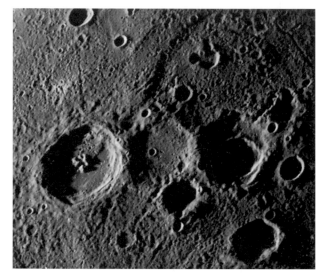

Thrust faulting produced this crater-deforming scarp east of Donne. NASA.

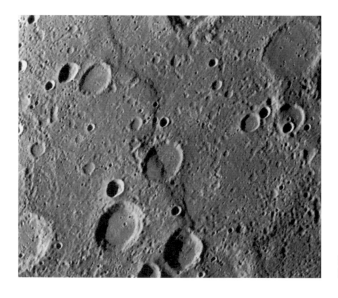

Part of the Santa Maria Rupes. NASA.

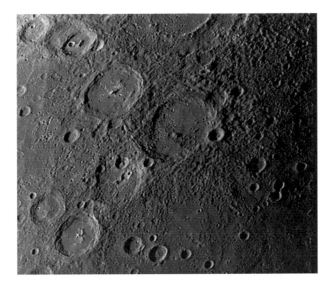
View of the Moliere region (Moliere at centre). NASA.

structuring radial to Rodin, noticeable in the rough terrain to its east. North of Rodin lies a large unnamed scarp and compressional ridge that runs parallel to Endeavour Rupes and Antoniadi Dorsum to its west. **Melville** (22N, 10W/154 km) looms at the limit of our current knowledge to the east of Rodin. Two large craters, **Molière** (16N, 17W/132 km) and **Abu Nuwas** (17N, 20W/116 km) adjoin Rodin's southern rim, while Rodin's northwestern wall is superimposed upon by **Ts'ai Wen-Chi** (23N, 22W/119 km), a younger crater with internally-terraced walls and a flooded floor which is cut through by an unnamed scarp to its north. Another large unnamed multi-ring basin, **Donne-Molière**, whose main rim measures some 1,060 km across, can be traced in this area. The crater **Donne** (3N, 14W/88 km) is superimposed upon the centre of this ancient basin. East of Donne is a remarkable chain of three similar-sized craters (all unnamed) which clearly demonstrates the principle of superposition; the western wall of an ancient crater (5N, 12W/50 km) is overlapped by a younger unnamed crater (40 km), whose western wall is itself overlapped by an even younger crater (50 km). Subsequent thrust faulting has produced a scarp to their north which cuts through the centre of the youngest crater. West of Donne is the **Santa Maria Rupes** (6N, 20W), an east-facing scarp some 2,000 m high that was caused by thrust faulting. Several craters in its vicinity have been deformed by the feature's formation, including one that has been divided down the centre and its opposing walls pushed closer together.

Moving across several hundred kilometres to the northeast of Santa Maria Rupes we come to **Sinan** (16N, 30W/147 km) and **Li Po** (17N, 35W/120 km), both ancient eroded craters with flooded floors. A dark, ancient lava plain extends west of the craters **Yeats** (9N, 35W/100 km) and **Handel** (3N, 34W/166 km), part of a more extensive dark plain (now largely overlain with craters and brighter ejecta) that extends for several hundred kilometres to the west, across to the small young impact crater **Tansen** (4N, 71W/34 km). Notable large craters in this area are **Proust** (20N, 47W/157 km), **Lermontov** (15N, 48W/152 km), **Giotto** (12N, 56W/150 km) and **Chaikovskij** (7N, 150W/165 km). Lermontov has a particularly bright interior, probably owing to being overlain with young ejecta material which originated from further afield. A prominent valley, **Haystack Vallis** (5N, 46W) radiates from the rim of a large unnamed crater on the equator and stretches some 200 km northwards to touch the eastern rim of Chaikovskij.

To complete this survey of the northeastern quadrant, we return to the area immediately west of the Mercurian *terra incognita*. The rim of the large and ancient crater **Vyasa** (48N, 81W/290 km) is overlain by two much younger craters – **Stravinsky** (51N, 74W/190 km) in the northeast and **Sholem Aleichem** (50N, 88W/200 km) in the northwest. Another large ancient and eroded crater (unnamed) occupies the area immediately north of Vyasa between Stravinsky and Sholem Aleichem – this may even be older than Vyasa itself, since the traces of Vyasa's northern rim can be traced superimposed upon the floor of this indistinct feature. A couple of hundred kilometres southwest of Vyasa is the crater

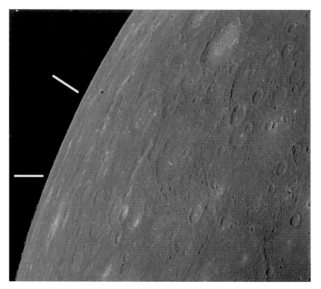

Oblique view of the Haystack Vallis (arrowed) area. NASA.

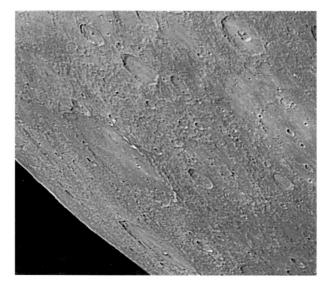

Oblique view of the Stravinsky region. NASA.

Al-Hamadhani (39N, 90W/186 km), a feature that has been overlain with wispy bright ejecta from further afield. Running between Vyasa and Al-Hamadhani is a bright linear feature (possibly an unnamed scarp) which runs southeast from around 48N, 105W to 41N, 85W. The region between Al-Hamadhani and Vivaldi is relatively smooth intercrater plains overlain with streaks of ejecta.

Other Named Craters in the Northeastern Quadrant (in order of Descending Latitude)

Myron (71N, 79W/31 km)
Al-Akhtal (59N, 97W/102 km)
Bruegel (50N, 108W/75 km)
Scarlatti (41N, 100W/129 km)
Sōseki (39N, 38W/90 km)
Mussorgskij (33N, 97W/125 km)
Vlaminck (28N, 13W/97 km)
Li Po (17N, 35W/120 km)
Aśvaghosa (10N, 21W/90 km)
Rajnis (5N, 96W/82 km)
Mistral (5N, 54W/110 km)
Al-Jāhiz (1N, 22W/91 km)

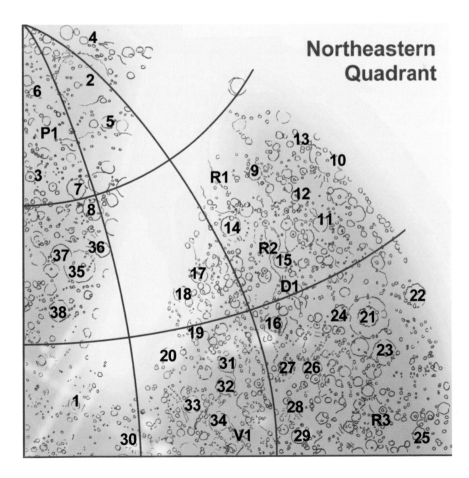

Northeastern Quadrant

Key to Map of Northeastern Quadrant

1. Vivaldi
2. Goethe
3. Gauguin
4. Aristoxenus
5. Tung Yüan
6. Despréz
7. Monteverdi
8. Rubens
9. Sor Juana
10. Monet
11. Gluck
12. Echegaray
13. Grieg
14. Derzhavin
15. Holbein
16. Wren
17. Hugo
18. Velázquez
19. Kuan Han-Ch'ing
20. Praxiteles
21. Rodin
22. Melville
23. Molière
24. Ts'ai Wen-Chi
25. Donne
26. Sinan
27. Li Po
28. Yeats
29. Handel
30. Tansen
31. Proust
32. Lermontov
33. Giotto
34. Chaikovskij
35. Vyasa
36. Stravinsky
37. Sholem Aleichem
38. Al-Hamadhani
D1. Antoniadi Dorsum
R1. Victoria Rupes
R2. Endeavour Rupes
R3. Santa Maria Rupes
P1. Borealis Planitia
V1. Haystack Vallis

Southeastern Quadrant

The Renior Region: *An Impressionistic Terrain*

The Southeastern Quadrant covers a large portion of Mercury's southern hemisphere, from the equator to the south pole, from longitude 10W to 100W. This area has a much more complete and more easily decipherable photographic coverage than the Northeastern Quadrant. Like the quadrant to its north, a sizeable proportion of intercrater plains covers the area. Islands of heavily cratered terrain crop up across the quadrant, and the region is riddled with scarps, ridges and valleys. Linear dustings of ejecta can be traced from numerous young impact craters.

Concentric to the equatorial Donne-Molière impact basin (see Northeastern Quadrant) are numerous (mainly indistinct) ridges that represent parts of the ancient feature's multiple rims. The clearest remnants of the Donne-Molière basin's rings can be found to the north of **Dvorák** (10S, 12W/82 km) including a section which is a continuation of the northern Santa Maria Rupes south of the tiny crater **Hun Kal** (1S, 20W/2 km). The centre of Hun Kal itself is used as a marker for the planet's 20° west meridian. The heavily cratered area around Hun Kal makes a striking contrast with a large smooth flooded area to the immediate south of **Lu Hsun** (0N, 23W/98 km), a crater with rounded walls and a striking valley cutting through its northeastern wall and floor. The young plain south of Lu Hsun fills an unnamed crater some 300 km in diameter; to its southeast lies **Brunelleschi** (9S, 22W, 134 km), a relatively well-preserved crater with wide, internally terraced walls and a broad outer apron to its west. Upon its smooth and flooded floor lie several ridges and a number of small impact craters, along with numerous peaks, including a well-formed linear massif just southeast of centre. Brunelleschi is superimposed upon the northwestern quadrant of an older, slightly smaller crater – an infrequently-seen clear-cut case (on any planet) of a large impact crater obscuring part of a smaller crater. Several rupes and valleys in this vicinity may represent part of the Donne-Molière basin topography. From the shadows evident in the area east of Dvorak, there are indications of another large (unnamed) impact basin whose centre lies at around 20S, 360W.

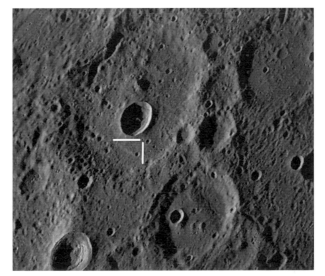

The diminutive crater Hun Kal (indicated), used as a marker for Mercury's 20° west meridian. NASA.

Towards the east, Dvorak is a lovely impact crater with a sharp rim, internally terraced walls and a sizeable central peak; its size, shape and setting in a heavily cratered region gives it a strong resemblance to the lunar crater Tycho (sadly, though, minus Tycho's magnificent bright ray system). A short distance to its south lies a similar (unnamed) impact crater, slightly smaller than Dvorak and having a small bowl-shaped impact crater embedded in its northern rim. This unnamed crater straddles the northern ramparts of a large unnamed crater some 220 km in diameter, on whose southwestern floor the whole of the crater **Hitomaro** (16S, 16W/107 km) resides. There is a stunning contrast between the ancient rubbly floor of the unnamed crater, which has traces of an inner ring and is striated with impact features radiating from Hitomaro, and the interior of Hitomaro itself, which is smooth and flooded, protruded through in the west by a fascinating complex of peaks.

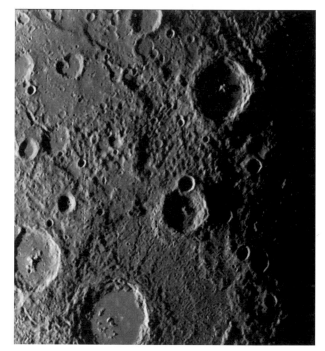

Dvorak (upper right) and Hitmaro (at bottom, left of centre). NASA.

A prominent conjoined trio of large craters punctuates Mercury's crust to the south of Hitomaro, namely (from west to east) **Mahler** (20 S, 19W/103 km), **Kenko** (22S, 16W/99 km) and **Balagtas** (23S, 14W, 98 km). Mahler is the youngest and best-preserved of the three, with its terraced walls, flooded floor and large central massif, while Kenko has a flooded, hilly floor but no prominent peaks. Balagtas is cut through from north to south by a tremendous scarp which may be connected with the Hiroshige-Mahler impact basin or, alternatively, the Ibsen-Petrarch basin. **Dario** (27S, 10W/151 km), a large crater with numerous elevations on its floor, lies at the very eastern edge of the quadrant; so too does Pigalle (39S, 10W/154 km) which lies several hundred kilometres south of Dario. Between the two craters is a broad section of hilly and lineated terrain which extends across to **Arecibo Vallis** (27S, 29W) in the west and **Vostok Rupes** (38S, 19W) in the south, covering some 75,000 square kilometres.

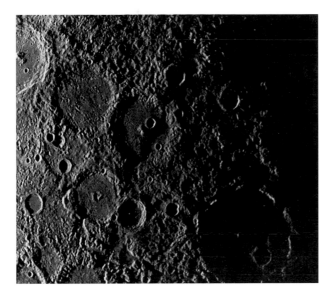

Balagtas (near centre) with Dario on the terminator (lower right). NASA.

Returning to the north of the quadrant once more, **Homer** (1S, 36W/314 km) is a large flooded impact basin that has retained half of an inner ring in an arcuate ridge. Interestingly, there rae strong indications that an area bordering Homer's southwestern wall may represent a pyroclastic deposit from an episode of volcanism; indeed, there appears to be a deep-seated fault feature in this location, a possible source of the eruption. A slightly larger (unnamed) impact basin adjoins Homer's western wall, from whose northern rim proceeds the mighty Haystack Vallis (see Northeastern Quadrant, above). This unnamed crater's western wall is grooved by a deep valley which runs south to **Rudaki** (4S, 51W/120 km). The crater **Titian** (4S, 42W/121 km) is located between Rudaki and Homer, and the whole area is underlain with a darker crust and peppered with small, bright impact craters.

One impact crater in particular – **Kuiper** (11S, 31W/62 km) – dominates the scene by virtue of its brightness and prominent ray system. Kuiper's spindly rays spread in all directions for several hundred kilometres, and can be traced most easily over the darker terrain to its west. Kuiper forms a dazzling diamond set in the ring of **Murasaki** (13S, 30W/130 km). Several interesting valleys radiate from the walls of Murasaki, all of them secondary impact features. One lies to the north, one to the southwest, another to the southeast, but the most prominent one, **Goldstone Vallis** (15S, 32W) cuts through the crust south of Murasaki for around 120 km. Adjoining Murasaki in the east is **Hiroshige** (13W, 27W/138 km), an older crater whose southern wall is host to an unnamed crater of similar size to Kuiper.

Southwest of Goldstone Vallis lies **Imhotep** (18S, 37W/159 km), an ancient, smooth-floored crater crossed by the rays of Kuiper. Imhotep is intruded upon in the northeast by a large unnamed crater which is traversed by a scarp; short sections of scarp also cross several other features to its east. The terrain to the west of Imhotep is ancient and moderately cratered, and crossed by Kuiper's rays. A large double-ringed flooded impact crater, **Renoir** (19S, 52W/246 km) dominates its environs. Some 300 km west of Renoir is **Repin** (19S, 63W/107 km), a somewhat younger crater with a prominent central peak and a smooth, flooded floor. Cutting across a larger unnamed crater to its north are several long fault valleys, including **Simeiz Vallis** (13S, 64W). A major unnamed valley, some 110 km long, is centred at 12S, 57W.

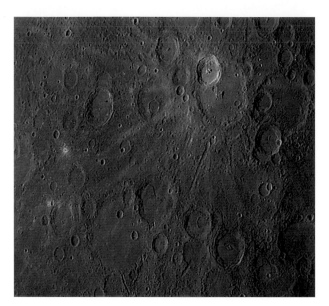

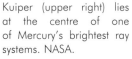
Kuiper (upper right) lies at the centre of one of Mercury's brightest ray systems. NASA.

Renoir and environs. NASA.

Petrarch (31S, 26W/171 km) is a large, smooth floored crater from whose northern rim emanates the deep Arecibo Vallis which cuts northwest through the Mercurian crust for nearly 100 km to join with the rim of a smaller flooded crater (unnamed). Arecibo Vallis is likely to be a graben, produced by crustal tension and faulting, rather than a secondary impact feature. South of Petrarch, several sizeable craters, including **Neumann** (37S, 35W/120 km), **Mofolo** (38S, 28W/114 km) and **Equiano** (40S, 31W/99 km) stand out. A number of medium-sized flooded craters lie scattered in the terrain west of Petrarch, and an unusual conjoined crater cluster is assembled around **Simonides** (29S, 45W/95 km), immediately west of which lies a small but brilliant double crater surrounded by a bright patch of rays (30S, 49W). Running across the landscape to the south is **Mirni Rupes** (37S, 40W), a notable scarp.

Petrarch and Arecibo Vallis (above centre) with Vostok Rupes (towards lower right). NASA.

One of Mercury's most prominent scarps – **Discovery Rupes** (54S, 38W) – cuts across the southern part of this quadrant, from the crater **Khansa** (60S, 52W/111 km) towards the western part of **Zarya Rupes** (42S, 22W), a total distance of nearly 1,000 km. In places the scarp – a thrust fault caused by crustal compression – exceeds a height of 2,000 m. The crater **Rameau** (55S, 38W/51 km) provides good evidence that Discovery Rupes is a compressional feature, since Rameau has been bisected by the scarp and its dimensions have been reduced perpendicular to the line of the scarp. Two smaller scarps, **Resolution Rupes** (62S, 52W) and **Adventure Rupes** (64S, 63W) can be found trailing beyond the southern point of Discovery Rupes, south of Khansa and **Rabelais** (61S, 62W/141 km) respectively. These scarps were probably formed at the same time as Discovery Rupes, also as a result of global crustal compression.

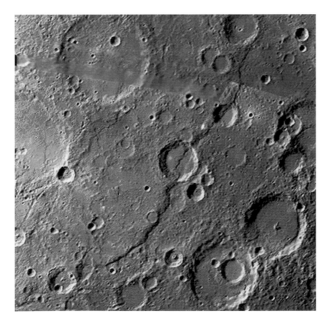

Discovery Rupes, an impressive compressional scarp, bisects the older crater Rameau. NASA.

A cratered region full of lineated valleys can be found to the east and southeast of Discovery Rupes. Noteworthy large craters in this region include **Sotatsu** (49S, 18W/165 km), a crater with terraced internal walls and a flooded, slightly convex floor; the battered and superposed **Kurosawa** (53S, 22W/159 km); **Hesiod** (59S, 35W/107 km) with its large internal crater and an unnamed crater adjoining its northern rim whose eastern floor margin appears to be grooved by a deep valley. Yet further south lies the very large crater **Pushkin** (66S, 22W/231 km) whose floor is flooded and traversed by several prominent wrinkle ridges. Embedded in Pushkin's northern ramparts can be found the younger impact crater **Tsurayuki** (63S, 21W/87 km) which intrudes upon **Mendes Pinto** (61S, 18W/214 km), an ancient crater with a coarse, hilly floor.

Finally, we return to the far northwestern part of the Southeastern Quadrant to complete our survey of this part of Mercury. Much of this region is covered with bright linear rays emanating from the relatively small craters **Snorri** (9S, 83W/19 km), **Copley** (38S, 85W/30 km) and an unnamed 20 km diameter crater to the west of **Chekhov** (36S, 62W/199 km). Chekhov itself is a multi-ringed crater, much like Renoir which lies 250 km to its north. Some 200 km to its southeast lies the large, smooth-floored crater **Schubert** (43S, 54W/185 km); a similar smooth-floored crater, **Haydn** (27S, 72W/270 km) is located on the opposite side of Chekhov. A larger, though less well-defined flooded crater, **Raphael** (20S, 76W/343 km) is one of this quadrant's largest named features. Its environs are made up of smooth intercrater plains, with little of particular note except the crater **Matisse** (24S, 90W/186 km), which lies some 260 km west of Raphael. To the north of Matisse, and completely enclosing the crater **Sullivan** (17S, 86W/145 km) lies an extensive patch of strongly ridged and lineated terrain with an area approaching some 50,000 square kilometres; it represents sculpting from the Tolstoj Basin. West of this area is a somewhat smoother unnamed plain, overlain with bright stripes of ejecta from young impact craters like Snorri.

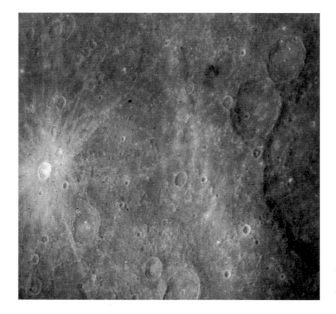

Snorri, a brilliant impact ray centre (at left) to the Simeiz Vallis (at right). NASA.

Two notable scarps can be found in this region. Around 500 km east of Copley is **Astrolabe Rupes** (42S, 71W), while some 800 km to Copley's south lies **Fram**

Rupes (57S, 94W); both are around 300 km long. Numerous smallish named craters populate the area between Rabelais and the planet's south pole; they include **Sei** (64S, 89W/113 km), **Camões** (71S, 70W/70 km), **Li Ch'ing-Chao** (77S, 73W/61 km) and **Sadi** (79S, 56W/68 km). **Boccaccio** (81S, 30W/142 km), a prominent, sharp-rimmed crater with a large central massif, is the most southerly named crater in this quadrant. A narrow linear valley bisects Boccaccio and proceeds some 150 km northwards in the direction of **Ovid** (70S, 23W/44 km). The terrain bounded in the triangle made by Boccaccio, Pushkin and Camões is relatively smooth and riddled with dozens of sizeable wrinkle ridges.

Mercury's south polar region, covering an area to 65 degrees south between 180W and 10E, from Leopardi (lower left) to Bach (top centre) and Ovid (lower right). NASA.

Other Named Craters in the Southeastern Quadrant (in order of Descending Latitude):

Polygnotus (0S, 68W/133 km)
Boethius (1S, 73W/129 km)
Lessing (29S, 90W/100 km)
Unkei (32S, 63W/123 km)
Rude (33S, 80W/75 km)
Carducci (37S, 90W/117 km)
Wergeland (38S, 57W/42 km)
Guido d'Arezzo (39S, 18W/66 km)
Nampeyo (41S, 50W/52 km)
Rilke (45S, 12W/86 km)
Po Ya (46S, 20W/103 km)
Sur Das (47S, 93W/132 km)
Bramante (48S, 62W/159 km)
Andal (48S, 38W/108 km)
Tintoretto (48S, 23W/92 km)
Ghiberti (48S, 80W/123 km)
Smetana (49S, 70W/190 km)
Africanus Horton (52S, 41W/135 km)
Shevchenko (54S, 47W/137 km)

Coleridge (56S, 67W/110 km)
Ma Chih-Yuan (60S, 78W/179 km)
Puccini (65S, 47W/70 km)
Callicrates (66S, 33W/70 km)
Holberg (67S, 61W/61 km)
Spitteler (69S, 62W/68 km)
Horace (69S, 52W / 58 km)
Okyo (69S, 76W/65 km)

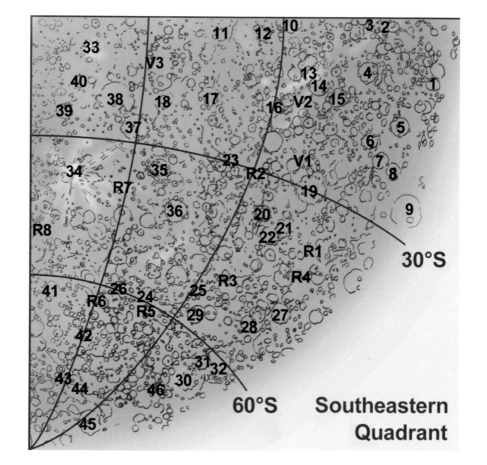

Key to Map of Southeastern Quadrant

1. Dvorák
2. Hun Kal
3. Lu Hsun
4. Brunelleschi
5. Hitomaro
6. Mahler
7. Kenko

8. Balagtas
9. Dario
10. Homer
11. Rudaki
12. Titian
13. Kuiper
14. Murasaki
15. Hiroshige
16. Imhotep
17. Renoir
18. Repin
19. Petrarch
20. Neumann
21. Mofolo
22. Equiano
23. Simonides
24. Khansa
25. Rameau
26. Rabelais
27. Sotatsu
28. Kurosawa
29. Hesiod
30. Pushkin
31. Tsurayuki
32. Mendes Pinto
33. Snorri
34. Copley
35. Chekhov
36. Schubert
37. Haydn
38. Raphael
39. Matisse
40. Sullivan
41. Sei
42. Camões
43. Li Ch'ing-Chao
44. Sadi
45. Boccaccio
46. Ovid
V1. Arecibo Vallis
V2. Goldstone Vallis
V3. Simeiz Vallis
R1. Vostok Rupes
R2. Mirni Rupes
R3. Discovery Rupes
R4. Zarya Rupes
R5. Resolution Rupes
R6. Adventure Rupes
R7. Astrolabe Rupes
R8. Fram Rupes

Northwestern Quadrant

The Shakespeare Region: *At the Edge of a Topographical Tempest*

The Northwestern Quadrant covers a large portion of the planet's northern hemisphere, from the north pole to the equator, from longitude 100W to 190W. Our current knowledge of much of the area through photographic coverage is reasonably thorough, given that the far western portions of the quadrant were imaged under low solar illumination, while features located more towards the centre of the disk show progressively less in the way of shadow and topographic detail.

Features centred around the Caloris Basin (whose centre actually lies a little beyond the area covered by this survey) dominate the Northwestern Quadrant, while a number of smaller multi-ring impact basins also occupy the region. No fewer than seven named plains spread onto the quadrant, namely: Borealis Planitia in the far northeast (see Northeastern Quadrant above); **Caloris Planitia** (30N, 195W)in the far west; **Suisei Planitia** (59N, 151W) in the northwest; **Sobkou Planitia** (40N, 130W) in the mid-east; **Odin Planitia** (23N, 172W) in the southwest; **Budh Planitia** (22N, 151W) in the mid-south and **Tir Planitia** (1N, 176W) in the far southwest of the quadrant.

With the exception of the region around Despréz (see Northeastern Quadrant) and the western parts of Borealis Planitia, the far northern latitudes of this quadrant are very heavily cratered. **Purcell** (81N, 147W/91 km), the most northerly named crater in this quadrant, occupies most of the floor of a larger unnamed crater. Purcell's floor is rough, with remnants of a central massif, and its southern wall is superimposed upon by a slightly smaller crater; beyond yet another unnamed crater further to its southwest lies **Van Dijck** (77N, 164W/105 km), a flooded crater whose floor is cut through by an arcuate east-facing scarp. Several larger, unnamed craters lie to the west; one noteworthy example is a particularly complex ancient crater that lies around 100 km to the northwest of Van Dijck at around 78N, 192W.

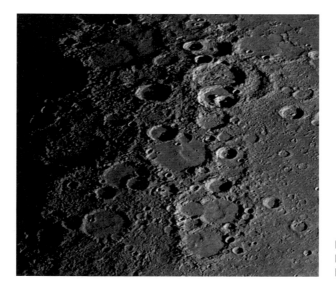

Near Mercury's north pole, Purcell (upper right) to Nizami (lower right). NASA.

Across the hilly terrain to the south of Purcell can be found a neat group of half a dozen medium-sized craters which includes the well-preserved **Jókai** (72N, 135W/106 km), the youngest and most prominent of the group, **Mansart** (73N, 119W/95 km) and **Bjornson** (73N, 109W/88 km). Each of the craters in this group has a well-defined rim, a flooded floor and remnants of a central elevation, that of Jókai being the most pronounced. Jókai is surrounded by a system of valleys which cut through its outer northern and eastern ramparts.

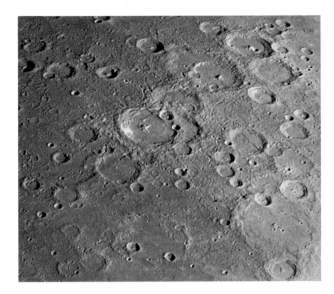

The group of large craters around Jokai. NASA.

Over the cratered plains south of Bjornson lies the ancient, eroded crater **Botticelli** (64N, 110W/143 km), notable for its patchwork floor of darker crustal material overlain with brighter patches of ejecta. Bright ejecta spreads mainly to the north, east and south from a small crater east of Botticelli on the 100° longitude. In the south, this mingles with other ejecta systems, including patches of material east of **Burns** (54N, 116W/45 km).

A sizeable area of this quadrant – around 250,000 square kilometres – is occupied by the smooth plain of Sobkou Planitia, which occupies part of the multi-ring Sobkou Basin, a feature whose main ring measures some 850 km in diameter. Sobkou Planitia is relatively free of scarps, ridges, fractures and valleys, although its southeastern edge is bordered by **Heemskerck Rupes** (26N, 125W), a scarp some 300 km long which interestingly follows along part of the line of a very broad, bright swathe around 1,000 km long and terminating just to the east of **Chong Ch'ol** (46N, 116W/162 km). The origin of this unusual bright feature is uncertain, since it doesn't appear to emanate from any particular impact crater; additionally, the area bordering the east of this bright line appears to be slightly darker than the surface further east.

One of Mercury's most prominent ray systems spreads across Sobkou Planitia from the conjoined craters **Brontë** (39N, 126W/63 km) and **Degas** (37N, 126W/60 km); Degas, the source of the rays, slightly overlaps the older Brontë. The brightest and longest component of the ray system stretches from the southeast of Degas, cutting at right angles across the unusual bright region mentioned above, across to the north of Vivaldi (see Northeastern Quadrant above) – a distance of around 1,000 km during which it gradually tapers from an initial width of around 30 km.

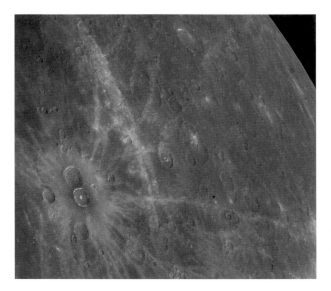

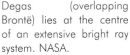

Degas (overlapping Brontë) lies at the centre of an extensive bright ray system. NASA.

Sprawling across more than 300,000 square kilometres, Suisei Planitia in the northwest is relatively smooth and considerably wrinkled with dorsa, many of which are related to the Brahms-Zola Basin underlying the western margin of the plain. Heavily cratered terrain lies to the north of Suisei Planitia, and is dominated by **Verdi** (65N, 169W/163 km) a prominent crater with strongly terraced walls, a large central massif and pronounced radial impact sculpting. An ancient flooded crater, **Turgenev** (66N, 135W/116 km) lies on the northeastern margin of Suisei Planitia, while further to the south lie the major craters **Ahmad Baba** (59N, 127W/127 km)

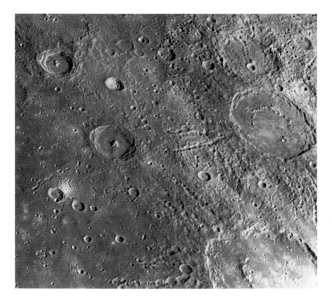

Strindberg (lower right) and Ahmad Baba (centre right) have prominent secondary radial cratering. Here the cratering converges on Kosho (left of centre). NASA.

and **Strindberg** (54N, 135W/190 km); both of these features display inner rings that partly protrude above their flooded floors, and both are surrounded by considerable secondary impact ridges, grooves and craters.

South of Suisei Planitia, the vast crater plain of **Shakespeare** (50N, 151W/370 km), Mercury's fifth largest named crater, melds with the smaller crater plain **Van Eyck** (43N, 159W/282 km) to form an imposing duo. **Mansur** (48N, 163W/100 km), a younger crater with an offset central mountain range, snuggles in the space west of Shakespeare, adjoining the northwestern wall of Van Eyck. From here, proceeding around part of the western periphery of Shakespeare are the **Zeehaen Rupes** (51N, 157W), an east-facing scarp some 200 km long. Two well-formed craters – **Zola** (50N, 177W/80 km) and **Brahms** (59N, 176W/96 km) stand guard over the western reaches of Suisei Planitia.

The great crater Shake-speare, at the centre of this image, is not easy to make out in this high-illumination image. NASA.

Odin Planitia and Budh Planitia, two medium-sized, but very different plains, lie side by side in the southern part of this quadrant. Odin Planitia, a grey hummocky plain, is one of the smoothest and least cratered areas on the entire planet. Unlike most other planitiae, Odin Planitia is not a flooded impact basin; it is an area smothered with thick blankets of debris splashed out by the Caloris impact. A number of circum-Caloris ridges wind their way across the plain, notably **Schiaparelli Dorsum** (23N, 164W), a broad ridge some 400 km long which is concentric with Caloris. **Couperin** (30N, 151W/80 km) and a ridge of mountains to its south marks the eastern border of an ancient, largely flooded unnamed Shakespeare-sized crater, and the eastern limit of Odin Planitia.

Immediately to the east lies the neighbouring plain Budh Planitia, which is somewhat darker than Odin Planitia. Traces of the multiple rings of the impact basin filled by the lava plains of Budh can be found most clearly towards the east, those in the west being disrupted somewhat by the system of ridges concentric to the Caloris Basin. The northeastern reaches of Budh Planitia blend with Sobkou

Schiaparelli Dorsum. NASA.

Planitia with little interruption save for a few hills and wrinkle ridges. In addition to numerous splodges and splashes of ejecta found across Budh, two small and particularly bright young unnamed craters shine out of the dark southern plain as a pair of glowing 'eyes' (around 16N, 156W), while buried in the southeastern plains are numerous medium-sized craters which show up as part-rings and ghost rings. Large craters of note lying beyond the plain in this area include **Harunobu** (15N, 141W/110 km), **Balzac** (10N, 144W/80 km), **Phidias** (9N, 149W/160 km) and **Tyagaraja** (4N, 148W/105 km). Both Balzac and Tyagaraja are relatively young impact features, with bright inner terraced walls, central massifs and radial impact sculpting, and both are surrounded by a dusky collar of impact melt; they make quite a stunning contrast with the large, smooth crater plain Phidias, which has a low, clear cut rim and patchwork interior.

Of all the known topographic features on Mercury, none is so extensive and has had so much impact on the rest of the planet as the giant multi-ringed Caloris Basin, which is centred just beyond the western limit of this quadrant. Caloris' main

The Caloris basin, Mercury's largest impact scar, shown here at left, its central plain straddling the terminator. NASA.

The area surrounding Amru Al-Qays (at centre). NASA.

ring, formed in part by the Caloris Montes, measures 1,340 across, and consists of segmented, smooth-surfaced massifs – great blocks of uplifted bedrock – which rise to heights of several thousand metres above the surrounding terrain. A gap several hundred kilometres wide can be found in the southeastern Caloris Montes, allowing the plains of western Odin to meet with the plains of Caloris. Traces of outer Caloris rings can be found among features much further east; the most distant structures concentric to the basin are traceable some 1,800 km from Caloris' centre.

Inside the Caloris Montes, the grey wrinkled plains of Caloris Planitia cover an area of almost 1.5 million square kilometres – somewhat larger than the Moon's biggest circular sea, Mare Imbrium. Concentric faults, scarps and dorsa vie for position on Caloris Planitia, along with a number of isolated peaks, small mountain clusters and small impact craters. Valleys and crater chains radial to the basin spread eastward, beyond the Caloris Montes and into the hummocky plains beyond. Radial features are particularly prominent in the region around the relatively young impact craters Zola and **Nervo** (43N, 179W/63 km) across to the terrain surrounding Van Eyck. Patches of lineated terrain radial to Caloris can also be found to the north and south of Odin Planitia and east of Tir Planitia. Much of this radial sculpting may be attributed to the low-angle impacts of material excavated by the Caloris impact, while some of it is also due to crustal movements along radial fault lines.

Superimposed upon the scarred terrain south of Caloris is the crater **Mozart** (8N, 191W/270 km), a large and imposing crater surrounded by an apron of radial impact sculpting and several coarse concentric ridges. A number of chains of secondary impact craters project in radial fingers away from Mozart, two particularly prominent ones lying to the southeast of the crater, running roughly parallel to one another across the western plains of Tir Planitia for around 250 km.

Tir Planitia itself is a broad, lava-flooded impact basin some 1,250 km in diameter which is divided into two by Mercury's equator; the basin's centre lies just above the equator at around 176W. Tir displays a number of Caloris-related topographic features, including a number of prominent ridges radial to Caloris and several patches of Caloris lineated terrain. Notable craters within this northern section of Tir include **Amru Al-Qays** (12N, 176W/50 km), a relatively young impact crater with terraced walls and an internal peak. A trio of slightly smaller unnamed impact craters with central peaks lie some 75 km to the east, south and west of Amru Al-Qays, and yet further beyond each of these craters can be found a young impact crater surrounded by a bright patch of ejecta.

Other Named Craters in the Northwestern Quadrant (in order of Descending Latitude):

Saikaku (73N, 176W/88 km)
Nizami (72N, 165W/76 km)
Martial (69N, 177W/51 km)
Kosho (60N, 138W/65 km)
Janácek (56N, 154W/47 km)
Whitman (41N, 110W/70 km)
Heine (33N, 124W/75 km)
March (31N, 176W/70 km)
Takanobu (31N, 108W/80 km)

Mickiewicz (24N, 103W/100 km)
Dürer (22N, 119W/180 km)
Chiang K'ui (14N, 103W/35 km)
Judah Ha-Levi (11N, 108W/80 km)
Wang Meng (9N, 104W/165 km)
Thoreau (6N, 132W/80 km)
Lysippus (1N, 133W/140 km)

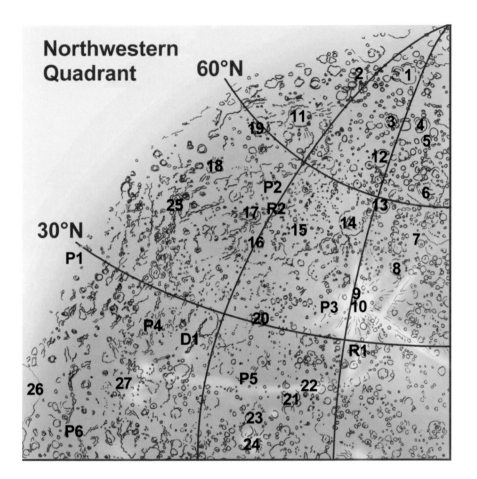

Key to Map of Northwestern Quadrant

1. Purcell
2. Van Dijck
3. Jókai
4. Mansart
5. Bjornson
6. Botticelli
7. Burns
8. Chong Ch'ol
9. Brontë

10. Degas
11. Verdi
12. Turgenev
13. Ahmad Baba
14. Strindberg
15. Shakespeare
16. Van Eyck
17. Mansur
18. Zola
19. Brahms
20. Couperin
21. Harunobu
22. Balzac
23. Phidias
24. Tyagaraja
25. Nervo
26. Mozart
27. Amru Al-Qays
P1. Caloris Planitia
P2. Suisei Planitia
P3. Sobkou Planitia
P4. Odin Planitia
P5. Budh Planitia
P6. Tir Planitia
R1. Heemskerck Rupes
R2. Zeehaen Rupes
D1. Schiaparelli Dorsum

Southwestern Quadrant

The Beethoven Region: *A Symphony of Splendours*

The Southwestern Quadrant covers the western portion of the planet's southern hemisphere, from the south pole to the equator, from longitude 100W to 190W. Like its neighbouring quadrant to the north, a reasonably thorough knowledge of this quadrant's topography has been gained through photographic coverage. The far western portions of the quadrant were imaged under a low angle of illumination from the Sun, while less topographic detail was mapped in areas located more towards the centre of the disk owing to the higher lighting conditions.

Several large, ancient multi-ring impact basins underlie much of this quadrant's topography. Northwestern regions are affected by the outer rings and lineations of the Caloris Basin. Straddling the equator are the Tir Basin in the northwest and the Mena-Theophanes Basin in the northeast. South of Tir lie the Tolstoj, Eitoku-Milton and the Vincente-Barma Basins, the outer rings of the latter meshing with the Bartók-Ives and the Hawthorne-Riemenschneider Basins. In the far south lies the circumpolar Sadi-Scopas Basin.

Beethoven (21S, 124W/643 km), the largest named crater on Mercury, dominates the northeastern portion of this quadrant. It is a clearly-defined impact basin with an underlying dark flooded floor that has been mottled with subsequent impacts.

Unlike many flooded basins, the floor of Beethoven does not display concentric wrinkle ridges, indicating that the underlying crust was sturdy enough to remain unflexed by the extra load of the infilling lava, so that the interior did not compress and slump to a great extent. Indeed, tensile forces appear to have been at work, causing an arcuate rille than runs across some 150 km of Beethoven's southern floor. Beethoven's centre is occupied by an unnamed 75 km diameter crater that is surrounded by a substantial outer rampart. A small but brilliant circular halo of ejecta surrounds an unnamed 10 km crater on Beethoven's northern floor. **Bello** (19S, 120W/129 km), a sizeable young crater with an internally terraced wall and bright central peak and surrounded by numerous radial ridges, sits on Beethoven's eastern floor. A north-south elongated flooded unnamed crater (80 × 60 km) lies on Beethoven's southeastern floor – its shape possibly resulting from a glancing asteroidal impact; a similar sized but more regular shaped unnamed flooded crater lies on Beethoven's southwestern floor. Beethoven's rim has a somewhat scalloped appearance owing to the embayment of numerous peripheral craters by lava flows. Beethoven's eastern and southeastern flanks in particular display prominent radial lineation which extends for more than 150 km eastward across the neighbouring plain – these features are radial to the Tolstoj impact basin. Beethoven's southern wall is straddled by the prominent crater **Sayat-Nova** (28S, 122W/158 km), a feature with internally terraced walls and a hummocky floor.

A sizeable smooth Calorian plain punctuated largely by old buried craters stretches between the equator and the northern ramparts of Beethoven. Bright, sharply-defined rays from the crater **Mena** (0S, 124W/52 km) cross the area, but a 45° arc break in the spread of the rays southwest of Mena forms a prominent and striking wedge shape. **Cézanne** (9S, 123W/75 km) lies on the plain between Mena and Beethoven, while several hundred kilometres further east can be found a remarkable line of half a dozen similar-sized craters (all unnamed), the final one of which lies immediately south of the crater **Chu Ta** (2S, 105W/110 km); three of these have sizeable central peaks, but the others have relatively smooth flat floors. It is perhaps a coincidence that these craters are radial to the Mena-Theophanes Basin, whose centre lies at 1S, 129W.

Philoxenus (9S, 112W/90 km), a large but unimpressive crater with a smooth flooded floor, is one of the few named features in the northeastern part of this quadrant. There are numerous similar flooded craters in this area, which opens out into a sizeable unnamed wrinkled plain further east (see description in Southeastern Quadrant, above). Bright spidery rays from a small unnamed impact crater (8S, 105W) radiate in all directions across several hundred kilometres. Bright ray systems also emanate from **Ives** (33S, 111W/20 km), some 450 km south of Beethoven's rim, and from **Schoenberg** (16S, 136W/29 km), just beyond Beethoven's western rim, the latter rays showing up particularly nicely against the dark Calorian smooth plains background.

Bartók (30S, 135W/112 km), a nicely-preserved impact crater with its bright, substantial terraced walls and central peak, makes a startling contrast with its much larger near-neighbour to the northwest, **Valmiki** (24S, 141W/221 km), a feature with low but clear-cut scalloped walls and a patchy flooded floor. There are indications of an inner ring to Valmiki, particularly in the west and east where small arcs of hills rise above the floor. The outer part of the crater's floor appears darker, perhaps due to later flooding along deep-seated fissures. Much of the local area is markedly striated with narrow valleys and catenae in a northwest-southeast pattern, features which are likely to be related to the Caloris or Tir impact basins.

Amid the hilly terrain northwest of Beethoven lie a number of old, somewhat eroded craters, including **Ts'ao Chan** (13S, 142W/110 km) and **Mark Twain** (11S, 138W/149 km). Of the two, Mark Twain is the more geologically interesting, since it displays a complete inner ring in the form of a low circular ridge – unusual for a crater of its relatively small dimensions. At 3S, 137W, some 170 km north of Mark Twain, there is an unnamed crater some 40 km in diameter, remarkable for its relatively huge central elevation; it possibly represents the largest central elevation in proportion to the size of any known Mercurian crater. It is rather similar in size and shape to the lunar crater Alpetragius, a crater near the eastern border of Mare Nubium; it measures 40 km across, has a floor some 3,900 m below its rim from which rises a rounded, 2,000 m high central peak one third the crater's diameter.

The southeastern portion of this quadrant includes a number of interesting large craters amid a patchy and varied background of lineaments, heavily cratered highlands and hilly terrain. **Michelangelo** (45S, 109W/216 km), with its near-complete inner ring of hills, is perhaps what Valmiki might have looked like before much of its interior was buried. Michelangelo's inner plain is particularly smooth; the plain between the crater's inner ring and its sharply delineated outer rim is less smooth, hillier in the northwest and covered in the southwest by secondary craters from nearby **Hawthorne** (51S, 115W/107 km). Beyond Michelangelo, a number of radial valleys, ridges and catenae stretch in all directions, meshing with similar patterns of terrain radiating from Hawthorne to the southwest. Hawthorne's central mountains form a tiny circlet of peaks a few kilometres in diameter rising from a very smooth floor. A curious broad meandering valley apparently composed of a series of linked craters, much like our own Moon's Vallis Rheita, lies just north of the crater's rim; considerably eroded, the feature likely represents a secondary impact feature related to one of the local ancient impact basins. **Hals** (55S, 115W/100 km), an older crater, lies beyond Hawthorne's southern ramparts. Several hundred kilometres across the small, smooth plain to the east of Hals lies **Riemenschneider** (53S, 100W), an ancient crater with a flooded floor. There are indications that the crater lies within a larger and more ancient basin more than 400 km across; the terrain in this vicinity is darkly mottled.

Another darkly mottled terrain – one that has clearly been sculpted by Calorian and Tolstojan impacts – surrounds the linked craters **Shelley** (48S, 128W/164 km) and **Delacroix** (45S, 130W/146 km). Shelley, the older of the pair, has rather eroded walls and a hilly floor whose interior has been streaked from the northwest with Calorian ejecta, although there remains an indication that the internal hills once had a ring structure. A prominent ridge extends into Shelley from the southeast, crossing its wall and intruding into its eastern floor. Bright rays emanating from **Han Kan** (72S, 144W/50 km), around 1,000 km to the south, adorn much of this quadrant, with a thin streak overlying Shelley's floor and another broader streak immediately beyond its western rim. Shelley's northern wall is superimposed upon by the southern rim of Delacroix, a crater with a central line of hills rising from a flooded floor which is crossed by rays from a fresh small impact crater perched on its southwestern rim. Straddling Delacroix's western wall is an unnamed 50 km crater with internal terracing and a cluster of central elevations. Two interesting features are to be found to in the vicinity of Delacroix. Curving through the hilly terrain for several hundred kilometres is an unnamed scarp (centred at 42S, 138W); this feature appears to be the southernmost component of a much larger scarp that runs for most of the way to Tolstoj, far to northwest. A prominent valley lies to the north of Delacroix, a linked chain of craters that is likely to be a radial feature of the Vincente-Barma Basin (52S, 162W).

Vincente (57S, 142W/98 km) is the most prominent crater in an area of several thousand square kilometres. It has a sharp rim and strong internal terracing, with a considerably slumped inner eastern wall. Its central uplift has been completely buried by impact melt, only to be replaced with a tiny central impact crater at some later date. Numerous chains of craters radiate from Vincente, a prominent bifurcated chain stretching across **Pourquoi-Pas Rupes** (centred at 58S, 156W) to the west, while the crater is itself crossed by two bright lines of ejecta from Han Can. A number of smaller craters (all unnamed) occupy the hilly region between Vincente and Hals, many of them with relatively smooth floors, and the region is crossed by four bright linear rays from Han Can. **Sibelius** (50S, 145W/90 km) lies to the north; it has terraced internal walls and a substantial central mountain. A small scarp runs from Sibelius' northern edge and skirts the eastern rims of a line of several smaller craters. Immediately to the east of Sibelius is a large unnamed crater (50S, 142W/100 km) which has been flooded and eroded.

Now we move across to the northwestern section of this quadrant and describe the area surrounding the major crater **Tolstoj** (16S, 164W/390 km), which is actually only the central flooded ring of the much larger Tolstoj multi-ring impact basin whose main ring consists of a discontinuous inward-facing scarp and measures 510 km across. The topographical effects of the Tolstoj impact can be seen in distinct units located across much of this quadrant and beyond. The main areas that retain clear Tolstojian sculpting lie immediately around Tolstoj itself, particularly to its east and south; the area surrounding the giant crater Shakespeare, particularly to

Shelley. NASA.

Vincente. NASA.

its east (see above); and the area surrounding Raphael (see Southeastern Quadrant above).

Tolstoj the crater is slightly polygonal in outline, and has a sharply-defined, somewhat scalloped rim featuring a number of flooded, embayed craters. Tolstoj's flooded floor has an area of around 120,000 square kilometres. It contains several sizeable near-completely flooded craters, along with a few younger craters that were formed after the last episode of flooding. Unlike the interior of Caloris, the floor of Tolstoj does not appear to have experienced any major slumping and compression; it is a smooth plain devoid of wrinkle ridges and concentric graben, indicating that the underlying crust was sturdy enough to remain unflexed by the lava infill, preventing the interior from compressing and slumping. There are clear indications of a smaller but largely buried internal Tolstojan ring some 260 km in diameter, the most easily traceable section of which lies parallel to the southeastern rim and consists of an inward-facing arcuate scarp some 150 km long. **Liszt** (16S, 168W/85 km), a crater with a smooth flooded floor, overlies the western wall of Tolstoj.

Blocky terrain lies between the inner and outer scarps of Tolstoj, and one of the most immediately noticeable features of this zone is the presence of a broad collar made up of diffuse patches of dusky terrain, averaging 125 km wide, that completely surrounds the crater. These patches are slightly bluer than the surrounding areas, suggesting a compositional difference. Pronounced radial lineations and grooves stretching for several hundred kilometres can be found beyond Tolstoj's outer ring; these were formed by secondary cratering followed by structural deformation. With its smooth floor, blocky inner region and strongly lineated outer zone, the Tolstoj Basin is in many ways like a miniature Caloris Basin, although Caloris displays a number of features which are not shown in Tolstoj (see Northwestern Quadrant above) owing to its larger size.

A cluster of half a dozen joined and overlapping craters, all of a similar size, is located 150 km to the north of Tolstoj's rim, the largest and most easterly member of which is **Po Chü-I** (7S, 165W/68 km), the only named crater in the area. Po

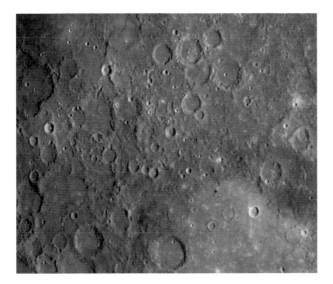

Tolstoj. NASA.

Chü-I is also the oldest of the group; it displays the most erosion and is overlapped by its western neighbour, which itself is overlapped by another crater to its north.

A more striking cluster of craters is to be found amid the strongly radially lineated ground to the northeast of Tolstoj. Dominating the group is **Zeami** (3S, 147W/120 km), a large and relatively young impact crater. It has a sharp, through somewhat irregular rim, marked internal terracing and an impact melt floor; a scattered group of bright central peaks rises from the floor, and these extend across Zeami's northern and western floor in two bright arcs which touch the crater's inner western wall. In contrast, the crater is surrounded by a prominent dark

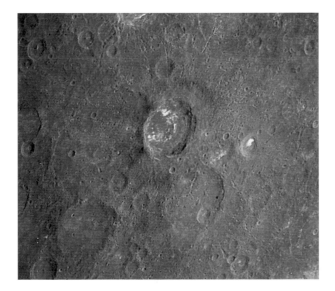

Zeami. NASA.

collar of impact melt, some 50 km wide, through which many radial chains of craters and deep secondary impact furrows extend into the surrounding terrain. Two particularly prominent continuous chains lie to the north of Zeami.

Theophanes (5S, 142W/45 km), a young impact crater notable for its brilliant central peak and northern floor, lies some distance to the east of Zeami. A small but prominent streak of bright ejecta nearby points towards**Sophocles** (7S, 146W/150 km), an older and considerably more eroded crater. Despite its age, indications of internal terracing remain within Sophocles; its northern floor is cut through by a curving chain of secondary craters thrown out by Zeami. Many small impact craters dot Sophocles' floor, along with a 35 km impact crater which nestles just within its southeastern wall. Due west, immersed within a landscape streaked with Tolstoj-radial furrows, is **Goya** (7S, 152W/135 km), a crater of similar age to Sophocles; low wrinkle ridges cross its floor. Goya abuts the southern rim of a very large, unnamed flooded impact basin (3S, 151W). Approximately 400 km in diameter, this basin is intruded upon in the east by Zeami and covered with that crater's ejecta and secondary craters; traces of a central elevation remain, in addition to a number of smaller flooded craters, and its western rim is clear-cut and easily traceable.

Tolstoj's eastern ramparts host **Rublev** (15S, 157W/132 km), a crater whose southern rim is highlighted by a bright young 20 km impact crater surrounded by a splash of light ejecta which extends across most of Rublev's floor. Secondary cratering and features radial to Rublev can be traced across the landscape, with the exception of the smooth plains of Tolstoj; particularly prominent radial features can be found to the north of Rublev.

Rublev. NASA.

Eitoku (22S, 157W/100 km) resides in a rougher neighbourhood southeast of Tolstoj. It has a low, rather scalloped rim and a very smooth floor; a sizeable group of peaks lies a little north of the crater's centre. Underlying Eitoku is a large unnamed flooded basin (centred at approximately 24S, 157W) in excess of 300 km across; ridges associated with the Tolstoj Basin run across the feature, so predates

Tolstoj by a considerable margin. The unnamed basin's floor is smoothest in the southwest, where a disjointed line of ridges delineates the main rim; numerous flooded craters and two large younger craters can be seen elsewhere across the floor. Furrows radial to the basin can be found in the vicinity, especially near Eitoku to its north and another to the southeast (centred at 26S, 153W).

Eitoku. NASA.

Heavily cratered terrain interspersed with patches of smooth plains and hilly terrain can be found in the region south of Tolstoj. A line of significant large craters stretches southward across the mid-southern latitudes of this area, beginning with **Basho** (33S, 170W/80 km), a prominent young impact crater surrounded by a dark collar of impact melt; it has a bright ray system, the most prominent of whose components stretches in a line to the northeast for several hundred kilometres over towards Eitoku. Basho's rim is sharp, and its bright internally terraced walls make a good contrast with its smooth dark floor; a large massif, several thousand metres high, rises at the crater's centre. In terms of its size, shape, setting and ejecta system, Basho resembles the famous lunar crater Tycho. North of Basho is an interesting unnamed crater (28S, 171W) identical to Basho in terms of size, terracing, smooth floor and large central massif, except that its southern floor is completely overlain by a younger crater and it is bisected by a ridge that may be associated with either the Tolstoj or Tir Basin. This ridge continues north of the crater for a further 150 km, extending across some hilly ground to the east of the large crater **Milton** (26S, 175W/186 km). Much of the area north of Basho is taken up by the ancient Eitoku-Milton Basin, whose centre lies midway between the rims of Milton and Tolstoj, although the feature is exceedingly indistinct from a topographical point of view.

A short distance east of Basho is **Ustad Isa** (32S, 165W/136 km), a smooth-floored crater covered with a pasting of light rays from its neighbour; sunk into its southern floor is an unnamed overlapping double crater (the larger component 40 km across), each component having a small central uplift which rises from a

Basho. NASA.

darker impact melt floor. Strongly Tolstoj-lineated terrain lies to the north of Ustad Isa. To Basho's immediate south is a large conjoined triple crater, while the hilly terrain to the west is markedly wrinkled, most of it probably associated with the Eitoku-Milton Basin.

A prominent trio of large craters comprising **Takayoshi** (38S, 163W/139 km), **Barma** (formerly known as 'Yakovlev': 41S, 163W/128 km) and a large unnamed crater (41S, 163W/120 km) lies on a moderately cratered plain midway between Tolstoj and Mercury's South Pole. Takayoshi has a smooth flooded floor punctuated by a few small craters and ridged with several sizeable wrinkles; such features are also found externally to the crater, particularly to its northeast, and they are

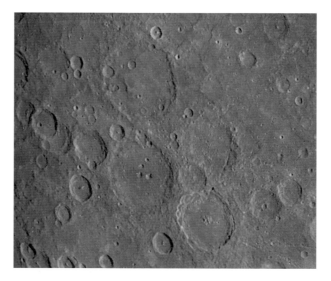

Barma. NASA.

probably related to the Eitoku-Milton Basin. An unnamed crater of similar age lies to the south of Takayoshi; its smooth floor is dotted with numerous secondary craters from Barma to its east, along with four sizeable keyhole-shaped craters which range from 5 to 15 km in diameter. Barma itself is the youngest of the trio, a typical medium-sized impact crater, with a sharp rim, particularly scalloped in the west, strong internal terracing and a large central massif rising from a smooth impact-fill floor.

Before venturing further south, we return to our circum-Tolstoj survey. Southwest of Tolstoj, in an area composed of a heady mixture of hilly and heavily cratered terrain which is riddled with valleys, scarps and ridges, lie a number of large craters worthy of note. **Hauptmann** (24S, 180W/120 km), a well-formed crater with a central massif, overlies the northeastern wall of a similar-sized crater. A sinuous valley runs around the far western floor of Hauptmann, cuts through the crater's southwestern wall and continues for more than 100 km through the terrain to the south. **Kalidasa** (18S, 179W/107 km) to the north is a fascinating impact crater with a strongly terraced inner western wall and a low but clear-cut eastern wall. A large, curving mountain massif separates the crater's smooth eastern floor from the hillier terrain making up the rest of the crater floor. Deep radial gouges caused by secondary impacts striate the landscape around Kalidasa; a group of around six prominent crater chains fan out beyond the crater's northern ramparts, while several more cut through the landscape to the southeast and south. Kalidasa itself is superimposed upon three older craters, the centres of each o which lie just outside of Kalidasa's rim; one to the east, one to the south and another to the west. Some fascinating topography lies further west, including an unnamed crater (21S, 185W/120 km) whose floor and southern wall appears to have been pulled apart by crustal tension.

Kalidasa. NASA.

Further north, the cratered landscape gives way to the broad, flooded impact basin of Tir Planitia. A little less than half of the area occupied by Tir Planitia (see also Northwestern Quadrant, above) covers the northwestern part of this quadrant.

Tir displays a number of Caloris-related topographic features, including a number of prominent ridges radial to Caloris and several patches of Caloris lineated terrain. Only one named feature stands out in this section of Tir – the little crater **Fet** (5S, 180W/24 km), a small but perfectly-formed impact crater which lies on a ridge pockmarked by dozens of small craters radial to Mozart (see Northwestern Quadrant above), some 500 km distant.

Dostoevskij (45S, 176W/411 km) is a little larger than Tolstoj; although it is Mercury's second biggest named crater, its ancient status renders it somewhat less imposing than many smaller features. Aeons of bombardment has worn away its walls, its floor has been filled with lava and overlain with ejecta from impacts further afield. Dostoevskij's western wall, in places overlain by smaller craters, is its most prominent feature; its eastern wall is less distinct, but those sections not obliterated by impact craters are visible as a blocky curves cut through by radial grooves, many of which extend into the plains beyond. One particularly long, well-preserved Dostoevskij-radial valley cuts through the terrain far to the southeast, extending from 53S, 167W to 59S, 157W – a distance of more than 300 km. An unnamed 100 km diameter crater occupies a large portion of Dostoevskij's northern floor, and much of the northern floor of this crater itself is taken up by a crater some 40 km across. Dostoevskij's southern floor is relatively smooth and unwrinkled, and it contains an interesting circlet of eight craters ranging between 5 and 20 km in diameter.

Fet. NASA.

It may not be an obvious topographic feature, but the large and very ancient Vincente-Barma Basin (see above) has affected parts of Dostoevskij's wall where the two features' rims intersect. The smooth flooded plain to the east of Dostoevskij is underlain by lava fill within the Vincente-Barma Basin, and some of the wrinkle ridges on this plain were produced post-fill. South of Dostoevskij lies **Dowland** (54S, 180W/100 km), a crater with broad terraced inner walls; it overlaps the southern wall of a smaller unnamed crater with a rather jumbled, blocky floor. A

Dostoevskij. NASA.

compressional lobate ridge, extending from the southwestern wall of Dostoevskij, cuts through the eastern part of the crater and proceeds another 100 km to the southeast. This is one of a number of dramatic northwest-southeast oriented ridges and rupes in the vicinity (see below).

Liang K'ai (40S, 183W/140 km), a large crater with a flooded floor, lies beyond Dostoevskij's northwestern rim. Of particular note is a large scarp some 400 km in length which bisects Liang K'ai and cuts through a number of craters to its north. **Gainsborough** (36N, 183W/100 km), a fairly unremarkable impact crater, lies nearby amid a heavily-cratered region which encompasses **Sarmiento** (30S, 188W/145 km), a crater at the current western limit of our knowledge of Mercury. The entire area east of Sarmiento, to the north and northeast of Gainsborough, appears to be strongly ridged in a pattern concentric with the Vincente-Barma Basin whose centre lies far to the southeast.

Several large scarps extend southeastwards across the terrain to the southeast of Dostoevsky into the hiller, more cratered Mercurian 'Antarctic circle'. One of the most prominent of these, **Hero Rupes** (centred at 58S, 171W), forms a great curved cliff some 700 km long. Although it is a lobate ridge caused by large-scale global tectonic activity and crustal crumpling, Hero Rupes appears to conform somewhat to the southwestern outline of the ancient Vincente-Barma Basin. A number of short ridges run parallel to parts of Hero Rupes, both in the lower ground to the west and in the higher terrain immediately to its east. Hero Rupes' main west-facing cliff begins east of Dowland inside an unnamed crater (55S, 175W/90 km) whose floor is cleft with an unusual system of spoke-like valleys radiating from a smaller crater at its centre. Hero Rupes completely bisects a chain of three craters to the south, goes on to reach its highest elevation of around 3,000 m ve the low smooth terrain to the west, before skirting the southern ramparts of an unnamed crater (62S, 167W/85 km) as it curves to the east. At its eastern end there a slight bifurcation, the lesser component sinking to the east while the main scarp proceeds southeast to finally terminate along the eastern rim of an unnamed crater (62S, 161W/30 km).

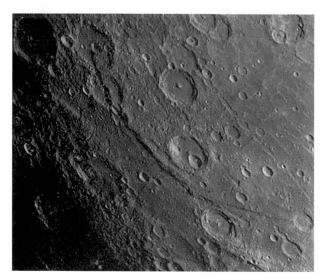

Formed by crustal compression, the mighty Hero Rupes is one of the terrestrial planets' largest cliffs. NASA.

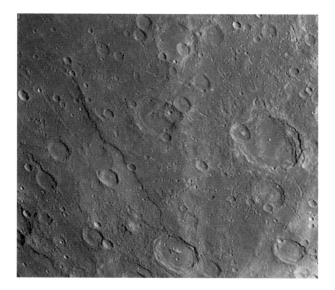

Pourquoi-Pas Rupes. NASA.

Another extensive lobate scarp, Pourquoi-Pas Rupes cuts across a tract of considerably less cratered terrain in a slight curve some 300 km long. Its main cliff faces west, as does that of Hero Rupes, although it only averages around half its height. Pourquoi-Pas Rupes is at its highest and widest in its northern section; mid-way it bisects an unnamed crater (58S, 156W/12 km), and it proceeds south to cut through the northern wall of a sizeable unnamed crater (60S, 155W/65 km), where it terminates in a sharp 90° turn to the west within that crater.

Gjöa Rupes (centred at 67S, 159W), the third named scarp in the area, curves for around 200 km across the cratered terrain between the southern end of Hero Rupes and the crater **Keats** (70S, 155W/115 km). Its northern section bisects a large unnamed crater (63S, 166W/110 km), branching into two on that crater's northern floor and cutting through the central massif of a smaller crater immediately to the south; it terminates in a low jagged edge to the north of Keats. Keats displays a substantial western wall and a tiny central elevation, but its other half is overlain by a 50 km crater whose eastern wall has been pushed to the west by crustal compression. To its south lies **Dickens** (73S, 153W/78 km), a younger crater with broad terraced inner walls and a central group of hills. An elevated tongue of ground lies adjacent to its western rim, which probably represents the vestiges of a crater upon which Dickens has been superimposed. West of Dickens can be found a very large unnamed crater (72S, 178W/200 km) whose southwestern rim is straddled by **Leopardi** (73S, 180W/72 km).

A group of large, prominent craters covers a large part of the southeastern portion of this quadrant. They include the joined duo of **Bach** (69S, 103W/214 km) and **Wagner** (67S, 114W/140 km), **Chopin** (65S, 123W/129 km) and **Alencar** (64S, 104W/120 km). Bach is one of a number of large flooded craters with inner rings to be found in this part of Mercury. Its outer rim is slightly out of circular, distorting towards Wagner to its west. The crater's inner ring is more or less complete and bounds a smooth flooded plain which is peppered with small impact craters and dusted with patches of ejecta. Hillier ground, consisting primarily of concentric features, lies between Bach's inner and outer walls. Its southeastern ramparts break down where they intersect with an adjacent large unnamed elliptical crater (72S, 96W/80 × 65 km), a curious formation with a very broad elevated spine running north-south through its major axis. Bach's southern floor adjacent its inner wall is trenched with a deep valley, possibly caused by the faulting as the floor infill slumped; a wrinkle ridge running adjacent to this valley to its north appears to

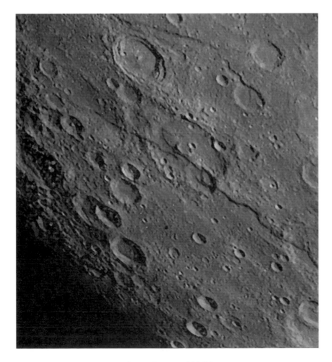

Gjoa Rupes. NASA.

Dickens crater (above centre), with Bernini (lower left) and Van Gogh (bottom centre). NASA.

support this scenario. Radial grooving can be found all around Bach, particularly to east and to the southwest, where a deep feature cuts through the northern part of **Cervantes** (75S, 122W/181 km).

Immediately to the west of Bach lies Wagner; a small crater obliterates the actual junction of both crater's rims. Although it is an old crater, Wagner retains a sharply pointed cluster of central hills. A young 45 km crater with a sizeable central peak overlies its southern rim, and from this a chain of three more similar-sized craters proceeds to the southeast; superimposed upon this group of craters are a couple of narrow linear crater chains which stretch some 250 km from the wall of Wagner to

Bach. NASA.

Belinskij (76S, 103W/70 km), although these features may be secondary to Belinskij rather than Wagner. Wagner's southwestern wall overlies an old and highly eroded unnamed 80 km crater (one of several ancient highly eroded craters of similar size to be found in the area) which is covered with secondary craters from the younger crater Chopin to the west. Chopin has a pronounced outline, with broad terraced inner walls and a large central massif that rises from a fairly smooth impact-melt floor. Ejecta and numerous deep linear crater valleys from Riemenschneider, which lies some 500 km to the northeast, extend across the terrain to the north of Chopin and Wagner.

Belinskij, a deep crater with a sharp rim, has been deformed by a large compressional ridge which runs through the plain south of Bach, from the rim of Cervantes to the plains near Camões, where it joins a number of similar ridges. The ridge widens to engulf the crater's western and northern ramparts, and several ridges proceed across the crater's floor around its central mountains. Cervantes to the west is a considerably eroded double-ringed basin which is overlain with ejecta, impact features radiating from several large nearby craters and a number of larger craters. Its northern rim is the most clearly defined component of the rings, while its southern rim is indented with a line of numerous deep secondary craters from nearby **Bernini** (79S, 137W/146 km). The northern part of Cervantes is cut through by a deep, slightly curving impact crater chain radial to Bach. Its inner ring, slightly northeast of centre to the outer ring, is battered and disjointed, overlain in the west by two 30 km craters and at its most complete in the south. Cervantes' central plain is smoother than the area between its inner and outer rings. Its northeastern and

Wagner. NASA.

southeastern walls are completely obliterated, while its western wall is straddled by **van Gogh** (77S, 135W/104 km).

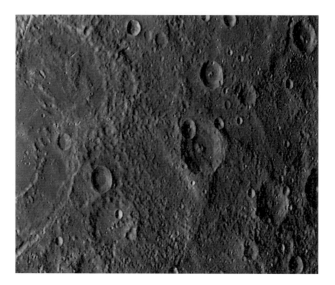

Belinskij. NASA.

Van Gogh, one of the youngest of the large craters, has a sharp rim, and its internally terraced walls surround an impact-melt floor. A collection of mountain peaks rises from the centre of the crater's floor, along with a large curving massif on the northern floor. Han Kan, one of Mercury's most notable young ray craters, stands out in the heavily cratered landscape north of van Gogh. Its sharp rim surrounds a broad, bright terraced inner wall and a prominent central mountain. The crater's floor appears to be streaked with bright ejecta, but the portion nearest the base of the southern wall appears very dark. A patchy collar of dark impact melt surrounds Han Kan, and many prominent bright linear rays streak away from the crater for many hundreds of kilometres (mainly to the north) to cover a large portion of the mid-southern section of this quadrant.

Abutting van Gogh's southern rim is the impressive double-ringed crater Bernini. This feature is rather deeper than other double-ringed craters in the region, and the inner walls of its outer ring are broad and terraced. Bernini's inner ring, some 70 km across, is comprises an intermittent series of hills arranged in a slightly elliptical and eccentric outline; a small mountain peak rises at the centre of Bernini's smooth central plain. The crater's northwestern wall is overlain by a young 50 km crater. Radial furrows and secondary chain craters extend around Bernini, but these are superimposed upon by a younger system secondary to the crater **Ictinus** (79S, 165W/119 km) some distance to the west. The largest of Ictinus' secondary impact valleys runs to the east, slightly curving around the southern ramparts of Bernini. Ictinus lies on a large curving scarp which extends north to Leopardi, another prominent scarp cuts across the terrain to the east of Leopardi. Leopardi's southwestern rim is indented by the large double crater **Scopas** (81S, 173W/105 km).

Mercury's South Pole itself is situated on the floor of **Chao Meng-Fu** (87S, 134W/167 km), a prominent crater with broad ramparts and a substantial central mountain complex. The terrain to its north is rough and peppered with hills and small craters, and represents the best-preserved section of the floor of the ancient and largely hidden Sadi-Scopas Basin.

Other Named Craters in the Southwestern Quadrant (in order of Descending Latitude):

Gogol (28S, 146W/87 km)
Surikov (37S, 125W/120 km)
Sibelius (50S, 145W/90 km)
Rimbaud (62S, 148W/85 km)
Yun Son-Do (73S, 109W/68 km)
Martí (76S, 165W/68 km)

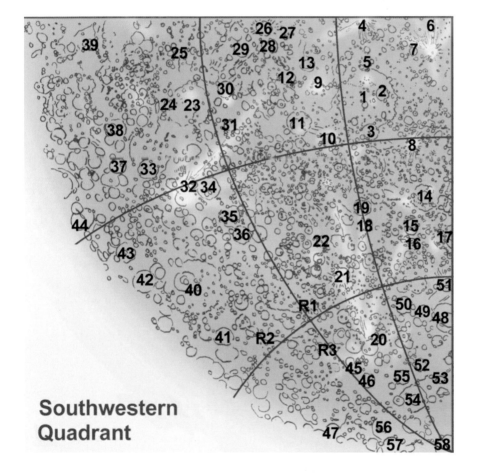

Southwestern
Quadrant

Key to Map of Southwestern Quadrant

1. Beethoven
2. Bello
3. Sayat-Nova
4. Mena

5. Cézanne
6. Chu Ta (crater chain)
7. Philoxenus
8. Ives
9. Schoenberg
10. Bartók
11. Valmiki
12. Ts'ao Chan
13. Mark Twain
14. Michelangelo
15. Hawthorne
16. Hals
17. Riemenschneider
18. Shelley
19. Delacroix
20. Han Kan
21. Vincente
22. Sibelius
23. Tolstoj
24. Liszt
25. Po Chü-I
26. Zeami
27. Theophanes
28. Sophocles
29. Goya
30. Rublev
31. Eitoku
32. Basho
33. Milton
34. Ustad Isa
35. Takayoshi
36. Barma
37. Hauptmann
38. Kalidasa
39. Fet
40. Dostoevskij
41. Dowland
42. Liang K'ai
43. Gainsborough
44. Sarmiento
45. Keats
46. Dickens
47. Leopardi
48. Bach
49. Wagner
50. Chopin
51. Alencar
52. Cervantes
53. Belinskij
54. Bernini
55. van Gogh

Chapter 3

Our Current Knowledge of Venus

Venus' Orbit

With an average distance of 108,208,926 km (0.7233 AU) from the Sun, Venus has a sidereal orbital period of 224.7 days. The planet's actual distance from the Sun ranges between 107,476,002 km (0.7184 AU) at perihelion and 108,941,849 km (0.7282 AU) at aphelion.

Inclined by 3.4° to the plane of the ecliptic (3.9° to the Sun's equator), Venus' orbit has an eccentricity of just 0.007, making it the most nearly circular orbit of any planet in the Solar System. Due to its almost circular orbit around the Sun, Venus' average orbital velocity of 36.020 km/s does not greatly differ from its maximum orbital velocity at perihelion (36.259 km/s) and its minimum orbital velocity at aphelion (35.784 km/s).

At its closest, Venus approaches the Earth to a minimum distance of 38,150,900 km (0.26 AU, or 2.1 light minutes) – the nearest approach to the Earth by any planet, and some 16.4 million kilometres closer than Mars at its closest perihelic approaches. At its furthest, Venus recedes to the other side of the Sun to a maximum a distance of 261,039,880 km.

Viewed from the Earth, the planet's synodic period (the period of time between two successive conjunctions with the Sun) is 583.92 days, a figure which is almost equivalent to five Venusian solar days. After eight Earth orbits and 13 Venus orbits, the two planets assume almost the same relative positions – just 0.032 percent away from a perfect orbital resonance of 8:13. Every eight Earth years and 13 Venus orbits around the Sun, Venus appears just 1.5° (about 22 hours) in advance of its former position. However close, this is considered to be an orbital frequency coincidence rather than a true orbital resonance between Venus and the Earth, since the slight difference adds up over time to produce a huge mismatch. The two planets are in exactly the opposite relative orientation after just 960 years. This affects the timing of transits of Venus, which occur in pairs separated by eight years, each recurring series of transits lasting 243 years.

A similar orbital frequency coincidence occurs between Venus and Mercury. In this case, the near resonance is in the ratio of 9:23. However, the mismatch is even greater than that between Venus and the Earth; Mercury is out of place by 4° each cycle, leading to a completely opposite relative orientation of the two planets in a period of just 200 years.

Physical Dimensions

Like all four of the Solar System's inner planets, Venus is a terrestrial planet composed chiefly of silicate rocks. Measuring 12,104 km across at the equator, Venus' diameter is only 650 km smaller than that of the Earth's equatorial diameter. The difference between Venus' polar and equatorial diameters is not known precisely, but it is thought to be very slight owing to the planet's slow rate of axial rotation; the difference is far smaller than the 43 km difference between the Earth's polar and equatorial diameters.

Encompassing some 938 billion cubic kilometres, Venus' volume is 0.857 of the Earth's. The planet's surface area amounts to some 460 million square kilometres, 0.9 that of the Earth – or, in more easily imaginable terms, equivalent to the area of the Earth minus the area of the North Atlantic Ocean.

As a consequence of its similar size and closeness to the Earth, Venus is sometimes referred to as our planet's 'twin' or 'sister' planet – although in many important respects the two planets could not be more dissimilar, leading some to jokingly refer to it as the Earth's 'evil twin'.

Mass, Density and Gravity

Like Mercury, Venus has no natural satellites of its own. Using Kepler's third law, analysis of the orbits of space probes around Venus has enabled allowed its mass to be determined at 4.87×10^{24} kg (4.87 sextillion tones/4,870 trillion tones), a mass 81.5% that of the Earth, making it the Solar System's seventh most massive object.

Venus' material has an average density of 5.204 g/cm^3 – slightly less dense than the mean density of either Mercury (5.43 g/cm^3) or the Earth (5.52 g/cm^3). Surface gravity at the planet's equator is 90 percent that of the Earth's, and its escape velocity is 10.46 km/s.

Axial Tilt and Rotation Period

Venus' rotational axis is tilted by a mere 2.64° to the plane of its orbit around the Sun, so there are negligible seasonal variations in the amount of sunlight and solar energy experienced by any particular part of the planet. The planet's north celestial pole lies at 18 h 10.9 m RA, 67° 9 minutes Declination – in the constellation of Draco, midway between the Draconian stars Nodus I and Altais. Chi Draconis, magnitude 3.56 and some 5.5° away, is the nearest bright star to Venus' north celestial pole; the planet's south celestial pole lies in the constellation Dorado, a few degrees away from the Large Magellanic Cloud. However, no stars are ever visible in normal light from Venus' surface owing to its perpetual blanket of dense cloud. Even the Sun itself, from Venus' surface some 44 arcminutes across and shining at an apparent magnitude of –27.6 is always completely obscured by cloud.

Venus has the slowest sidereal rotation rate of any of the planets, turning once on its axis every 243.02 days. The planet's rotation is retrograde – when viewed from above the north pole, it turns in an anticlockwise direction – a phenomenon shared only by Uranus among the major planets. A sidereal day on Venus lasts 18.3 days more than its sidereal orbital period around the Sun. From the surface of

Venus at a point on the equator the Sun (if, of course, the Sun were ever visible) would appear to rise in the west (taking 4 ¾ hours to heave itself completely up over the horizon) and sets in the east some 116.75 days later (taking an equally long time between first touching the horizon and disappearing below it). At the equator, Venus' surface rotates at a velocity of just 6.5 km/hour – around average walking speed, considerably slower than a point on Mercury's equator and 246 times slower than a point on the Earth's equator.

Origins

Four large protoplanets composed largely of silicates and metals – Mercury, Venus, Earth and Mars – had grown to dominate the inner Solar System within around 200 million years of the collapse of the solar globule and the formation of the protosun. None of these protoplanets is thought to have been massive enough to have attracted a disk of material from which its own satellite could have formed.

The protoplanets of the inner Solar System swept up all the available material within their own orbits, attracting dust and gas and accumulating large chunks of solid matter through countless impacts. High temperatures within the planet came about as a result of asteroidal impacts, internal pressure and the radioactive decay of elements, and as a consequence the material of the protoplanets melted. Differentiation took place as heavier elements sank to form planetary cores and lighter material rose to form their mantles and crusts.

Venus' current slow retrograde rotation period is dramatically different to the much faster prograde rotation period that it (along with all the other major planets) is likely to have had when it was a young planet, freshly formed out of the solar nebula. Current models of the dynamics of the early Solar System show that the angular momenta of the planets and their orbits are in the same direction as the initial angular momentum of the solar nebula. A planet's rate of rotation can change because of tidal gravitational interactions between it, the Sun, any satellites it may have and to a small extent other planets. In the case of Venus, tidal effects produced by the Sun on its dense atmosphere have undoubtedly had a braking effect on its rotation, but simple Sun-Venus tidal interactions fail to account for the planet's retrograde rotation. It has been suggested that a huge jolt to the system took place early in the planet's history when a smaller protoplanet collided with Venus, drastically reversing the planet's direction of rotation. There's no direct evidence for such an impact, as found in the case of the Earth's Moon, which is thought to have resulted from the impact of a Mars-sized planet some 4.6 billion years ago. If Venus did once form a satellite through similar means, it has long since disappeared, torn apart by tidal forces, the fragments being distributed in the inner Solar System or laid waste by impacting on Venus' surface.

Surface History

As the protoplanetary crusts thickened and consolidated they began to retain the imprints of countless asteroidal impacts in the form of craters and basins. Lava flows intruded through crustal fissures, filling the floors of many of the impact features. Clearly visible remnants of an intense period of asteroidal impacts called the 'great bombardment', which ended about 3.8 billion years ago, can be seen on the Moon, Mercury and Mars. Some of the impactors probably came from the inner Solar System, but many originated in the outer Solar System, diverted by gravitational interactions with the giant outer planets.

Yet no trace of this ancient period of high bombardment rates remains to be seen on either Venus or the Earth. Although nearly 1,000 impact craters have on been identified Venus – around four times the number that have been identified on the Earth – these features are very young in comparison to most of the large craters on the Moon, Mercury and Mars. A constant process of resurfacing by volcanic means has obliterated all the ancient Venusian impact features. It has been estimated that the average age of Venus' surface is just 500 million years old. The difference in the numbers of impact craters found on the Earth and Venus – two similar-sized planets – is explained by several factors. Seventy five percent of the Earth's surface is water overlying oceanic crust, and the sea floor is in a constant state of renewal by sea floor spreading. For example, the oldest rocks found on the Atlantic floor are just 180 million years old; everything older has been shunted sideways from the mid-ocean ridges and subducted beneath the continents on either side of the Atlantic. Erosion, sedimentation, plate tectonics and crustal crumpling has helped obliterate the most ancient impact features on the Earth, leaving behind a just few of the younger examples.

Although spreading mid-oceanic type ridges, transverse fractures and subduction zones aren't in evidence anywhere on Venus, that is not to say that there has been a lack of crustal movement and tectonic activity on the planet. Much of Venus' surface geology displays a strong resemblance to that found in terrestrial continents, with numerous large mountain ranges, zones of crustal rifting and major strike-slip systems where faulting has produced a relative horizontal movement of crustal blocks.

Tectonic Features

Like the Earth, Venus has a substantial molten nickel-iron core, a hot, ductile rocky mantle and a rocky crust. Its core is thought to be about the same size as the Earth's, some 2,400 km across. Venus' crust, composed of basaltic-type material similar to the material making up the Earth's oceanic plates, ranges between 20 and 40 km thick; Earth's oceanic crust is some 10 km thick, while its continental crust has an average thickness of 40 km. For decades, planetary scientists have attempted to understand how two planets of such similar size, structure and composition came to have such different crustal features.

It appears that Venus' global tectonics are driven almost exclusively by plumes of hot material arising deep within the mantle and intruding into the lithosphere, producing volcanic uplifts and volcanically active areas known as hot spots. Hot spots are known on the Earth – one of the best-known examples are the Hawaiian islands in the Pacific Ocean, which, because of sea floor movement over a relatively static hot spot, has created a chain of elevated features, the youngest of which is still volcanically active.

Venus does not have an active, mobile oceanic-type crust like the Earth, and its continental plateaux are not in the process of drifting. On Venus, the convective movement of the mantle at its hot spots produces deformation of the crust above, producing both compressive and tensile features on a regional scale. Such features include faults and wrinkle ridges, many of which occur in parallel or run in the same direction in accordance with the direction of deformation and crustal stresses.

Another form of tectonic feature known as tesserated terrain (from the Greek word 'tessera', a mosaic tile) is unique to Venus, and covers ten percent of the planet's surface. The most tectonically altered form of terrain on Venus, tesserated terrain consists of hilly sections of crust which have been fractured by networks of criss-crossing faults (usually trending towards one particular direction) into smaller angular blocks ranging from a few kilometres to more than 20 km across. Tesserated terrain is thought to have chiefly arisen as a result of crustal compression through tectonic activity, and represents the oldest crustal unit in any particular setting. It is possible that the crust in these areas was originally weaker than surrounding areas, perhaps made up of a different type of rock and more easily prone to faulting and deformation. Alternatively, tesserated terrain may underlie a significantly larger proportion of Venus' surface, those tesserae that are visible representing just a small sample of ancient crust that was largely buried by lava flows during the last major phase of planetary resurfacing.

Close-up of a typical patch of tesserated terrain. NASA.

Terrae and Highlands

Over time scales of hundreds of millions of years, Venusian hot spots produce a series of characteristics in the overlying crust. First, a convective plume in Venus' mantle creates a large uplifted region or dome. Stresses in the uplifted crust produce faults, rifts and huge valleys called chasma. The hot intrusion melts the surrounding rocks and volcanic activity breaks out onto the surface, creating volcanoes and depositing large quantities of lava. Over time, the crust thickens further as a volcanic plateau is built up over the hot spot.

Venus has three very extensive highland areas — Ishtar Terra, Aphrodite Terra and Lada Terra – each of which is of comparable size to the Earth's continents, plus a smaller highland plateau called Beta Regio.

Ishtar Terra (centred at 70N, 28E) dominates the planet's northern hemisphere. Measuring some 5,600 km from east to west, this vast highland plateau has an area of around 8 million square kilometres, equivalent to that of Australia. Lakshmi Planum (centred at 69N, 339E), a vast plain some 2,345 km across, occupies a sizeable portion of eastern Ishtar Terra. Lakshmi Planum is edged by four vast mountain ranges, namely Maxwell Montes in the east, Frejya Montes to the north, Akna Montes to the west and Danu Montes to the south. Of these ranges, the peaks of Maxwell Montes reach the highest altitude – up to 11 km. Lakshmi Planum and its surrounding mountain ranges is reminiscent of the Earth's Tibetan Plateau and its mountain borders. The two features probably share the same mode of origin – both features are converging plate boundaries which have produced a folded mountainous uplift. Lakshmi Planum is twice as large as the Tibetan plateau.

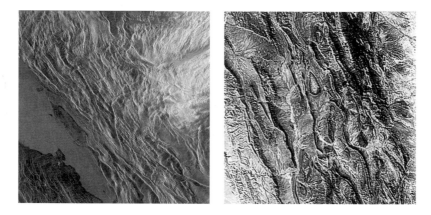

At left, Venus' Akna Montes (centred at 69N, 319E), a large mountain belt formed by crustal folding under northwest-southeast compressional forces. At right is a similar feature on the Earth, the Atlas Mountains in North Africa, which formed as the African plate crumpled into the southern margin of the European plate. NASA.

Aphrodite Terra (centred at 6S, 105E) is some 10,000 km long and stretches almost one-third of the way around Venus, most of it lying south of the equator. It has an area of around 15 million square kilometres (around half the size of Africa). Much of Aphrodite Terra comprises rough terrain that reaches heights of up to 7 km. There's plenty of evidence that crustal forces coming from more than one direction have been at work in the region, producing complex folding, ridges and fracturing. It is possible that this is a region of tectonic tension and new crustal formation, akin to terrestrial ocean floor spreading about a mid-ocean ridge.

Lada Terra (centred at 63S, 20E) is around 8,600 km across and is the least precipitous of Venus' highland continents. It contains much tesserated terrain and several large coronae; a number of deep graben-type canyons can be found cutting across the continent's northern part.

Measuring some 2,869 km in diameter, Beta Regio (25N, 283E) is one of the largest examples of a Venusian volcanic rise. Theia Mons (23N, 281E), a large shield volcano at the centre of Beta Regio, measures 226 kilometres across at its base and

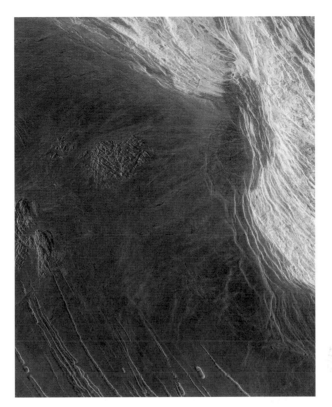

The eastern edge of Lakshmi Planum and the western edge of Maxwell Montes. NASA.

rises to a height of more than 4,500 m; a sizeable caldera measuring 50 × 75 km marks its summit. The volcano is surrounded by an extensive field of lava flows. Theia Mons lies at the centre of the system of rift valleys traversing Beta Regio, many of which run radially to the volcano. These graben typically measure between 40 and 160 km long, 40 and 60 km in width and often have uplifted margins

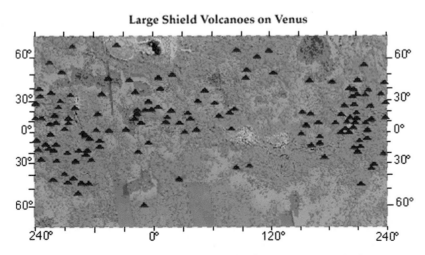

Large Shield Volcanoes on Venus

The global distribution of large shield volcanoes on Venus. NASA.

between 500 and 1,000 m high. Analysis suggests that they were formed as a result of some 5 to 10 percent crustal extension as the crust was put under tension during the uplifting of Beta Regio. A number of these graben extend beyond Beta Regio and cross the surrounding lowlands.

Venus, the Real Vulcan

Venus is the most volcano-ridden planet in the entire Solar System. Distributed within several hundred large volcanic fields on Venus are more than 55,000 volcanoes whose bases are bigger than 1 km in diameter; the total number of volcanoes exceeds 100,000 and may well be nearer the one million mark. By far the majority of them assume either large or small shield form, and while it is highly probable that volcanism still takes place on Venus, it is not known how many volcanoes are currently active. In comparison, the Earth's land surface has around 1,500 active volcanoes, most of these being above subduction zones like the 'ring of fire' encircling the Pacific Ocean, while there are probably around five times as many active submarine volcanoes to be found along mid-ocean ridges.

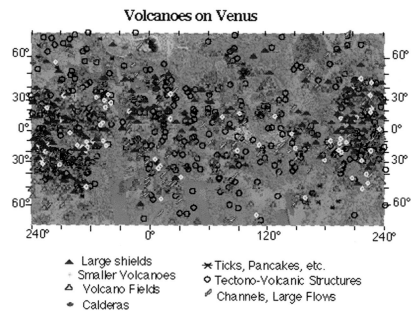

The global distribution of volcanic features on Venus. NASA.

Shield Volcanoes

Venus' surface boasts more than 150 large shield volcanoes, with bases ranging between 100 and 700 km across and heights of between 300 and 5,500 m. In comparison, the Earth's largest shield volcano, Mauna Loa, measures 120 km across at its base and rises to a height of 8,000 m. Venusian shield volcanoes are scattered across most parts of the planet and there is very little evidence of chaining, unlike many volcanoes found on the Earth. The lowland plains and highland regions contain relatively few large volcanoes; instead, a disproportionate number can be found in upland regions which are between 2 and 3 km higher than the global average.

Although Venus' shield volcanoes are generally broader and flatter then those found on the Earth, they resemble their terrestrial counterparts in that their gentle slopes are covered by long, radial lava flows radiating from a central vent or summit caldera.

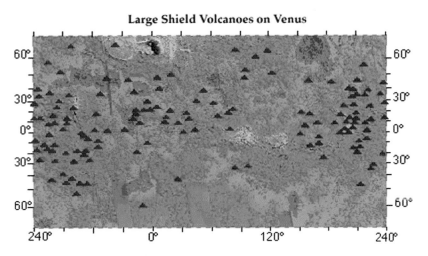

The global distribution of large shield volcanoes on Venus. Each red triangle marks the site of a shield volcano over 100 km in size. NASA.

Some notable large volcanic shields on Venus

Name	Location	Base dia (km)	Height above surroundings (metres)	Caldera dia (km)
Theia Mons	23N, 281E	226	4,500	75 × 50
Sif Mons	22N, 352E	300	2,000	50 × 40
Gula Mons	22N, 359E	276	3,000	40 × 30
Sapas Mons	9N, 188E	217	1,500	25
Ushas Mons	24S, 325E	413	2,000	15

Venusian volcanism displays a muted eruptive style involving fluid lava flows that emanate from a central caldera or along fissures, much like the type of eruptions that take place in Hawaii's Mauna Loa volcano, for example. There is little evidence of explosive, ash-forming eruptions of the terrestrial Vulcanian or Vesuvian type on Venus. There is however a great deal of evidence of viscous, silicate-rich lava flows. There are many examples of small rounded 'pancake' domes with summit vents, flattened tops and steep edges, built up by viscous lava and/or

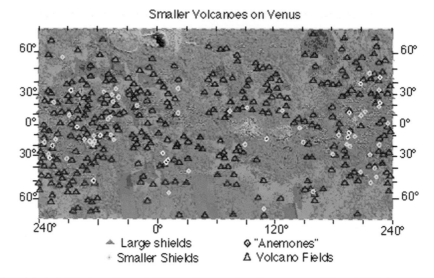

Smaller Volcanoes on Venus

Large shields   ◇ "Anemones"
Smaller Shields   △ Volcano Fields

The global distribution of small shield volcanoes (ranging between 20 and 100 km across), small shield fields and volcanoes with 'anemone' (flower-like) patterns of lava flow. NASA.

ash deposits. Usually associated with areas rich in faulting and crustal deformation, such as that found around coronae and tesserae, these 'pancake' domes are typically between one and two orders of magnitude larger than similar features found on the Earth.

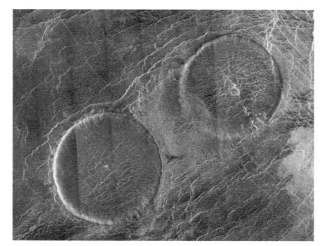

Volcanic 'pancake' domes in Tinatin Planitia, located at 15N, 9E. The largest has a base 62 km in diameter. They were formed by highly viscous lava, hence their flattened tops and steep margins. NASA.

Numerous factors have constrained explosive Venusian eruptions. The planet's high atmospheric pressure dictates that erupting lavas must contain a far higher gas content than their terrestrial counterparts to produce an explosive eruption at the surface. Moreover, terrestrial explosive eruptions rely largely upon a high water content within the rising magma and in the intruded rock, an uncommon substance within the Venusian mantle and crust. On the Earth, major volcanic explosions called phreatic eruptions take place when hot rising magma (between 600 and 1,170°C) comes into contact with ground or surface water; the water is almost instantaneously heated to steam, producing an explosion which can throw out vast quantities of ash, rock and volcanic bombs. A notable terrestrial example of this type of eruption ripped the top off Mount St Helens in the United States in 1980.

Only a small proportion of Venus' smaller volcanoes lie within the highland regions; most can be found in the lower elevations and on the plains. Some 80 percent of Venus' lowland plains are composed of volcanic lava flows; many of these flows undoubtedly cover large numbers of smaller volcanoes.

Most of Venus' estimated 100,000+ volcanic shields are small, with base diameters in the order of 30 to 40 km. Among this vast number, a couple of dozen are known as 'anemone' shields. Peculiar to Venus, these features consist of a central vent surrounded by flower-like patterns of narrow lava flows which, to mapping radar, appear as bright streaks encircling the volcano. Another unusual form of Venusian volcano, known as the 'tick' volcano (owing to its resemblance to that tiny insect) consists of a smooth depression containing a caldera, encircled by a raised rim and a surrounded by a number of radial ridges. Evidence that many volcanic domes have collapsed at their margins can be seen in those domes which have aprons of landslide debris, scalloped edges and/or concentric fractures.

An example of a Venusian 'anemone' shield, an unnamed feature measuring 40 km in diameter and located at 10S, 201E. NASA.

Many examples of volcanic cones, similar to those found on the Earth and Moon, can be seen across Venus. Venusian cones are mainly circular in outline with steep slopes that range between 12 and 23°C and heights of between 200 and 1,700 m. Remnants of the original volcanic vent are nearly always seen at their summits, but individual lava flows are not usually visible on the cone's slopes. Generally occurring in small groups clusters on fractured plains, some cones show evidence that they were formed before the crust experienced faulting (being cut through by faults), while others were clearly formed after faulting took place (superimposed upon the faults).

Lava Flows

Lava flows cover much of the surface of Venus. There are three types of lava flow feature: large lava flow fields (called 'fluctus'), long lava channels and their associated volcanic depressions.

Large Flow Fields

Venus has more than 50 large lava flow fields, most of which range between 100 and 700 km in length and average some 50 to 300 km in width. Most of the flow fields are found around the elevated borders of the lowland plains and feed into the plains, but some can be found in close proximity to volcanic shields. One of the largest of Venus' lava flow fields, Mylitta Fluctus (56S, 354E) is 1,250 km long and 500 km wide and covers an area 400 times that of the largest basaltic lava field in the United States (Craters of the Moon National Monument in Idaho), or roughly equal to the area of the basaltic lava sea of the Moon's Mare Imbrium. Like many of Venus' large flow fields, there's ample evidence for a series of eruptions, most originating at a single source, over a period of many years.

In some cases, such as can be found in the the flat plains of eastern Lakshmi, the nature of the flows may be deduced from its degree of radar reflectivity. Dark flows are likely to indicate smooth lava flows of highly ductile pahoehoe-type basaltic lava; brighter flows are likely to have somewhat rougher surfaces, similar to the slow-moving aa lava flows. Some lava flows display well-defined streamlined islands of elevated ground, shaped like teardrops in accordance with the direction of flow around them.

Lava Channels

Formed by the action of flowing lava, and superficially appearing like dried up river beds, the lava channels on Venus closely resemble sinuous rilles found on the Moon. Around 200 lava channels have been identified, most of which meander their way through lava flood fields or trickle down the sides of volcanic shields. They are often found in groups, like their lunar counterparts, and like lunar sinuous rilles they average between 500 m to 1.5 km in width. Their lengths range between a few tens of kilometres to several hundred kilometres, most being shorter

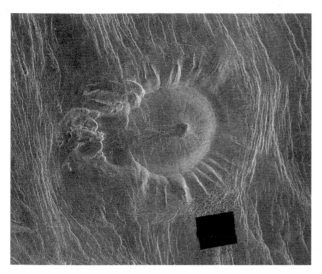

An example of a Venusian 'tick' volcano, an unnamed feature measuring 30 km across and located at 20S, 3E. NASA.

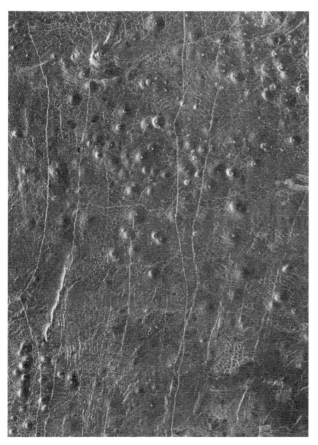

A sizeable field of volcanic cones lies on the fractured plain of Niobe Planitia. These cones average 2 km wide and rise to heights of around 200 m. NASA.

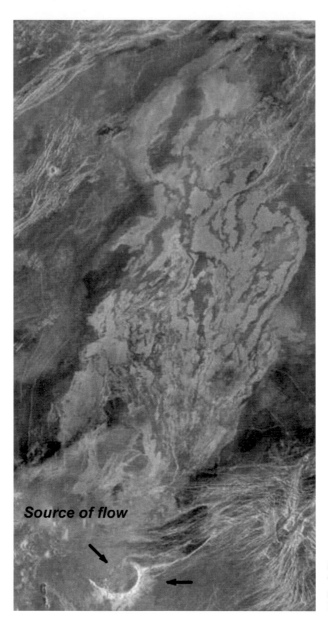

Source of flow

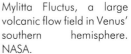

Mylitta Fluctus, a large volcanic flow field in Venus' southern hemisphere. NASA.

than 400 km. Baltis Vallis (centred at 37N, 161E) is the longest of Venus' lava channels – indeed, at 6,800 km long it is the longest lava channel in the entire Solar System. It traces a path across the surface to the west of Atla Regio northward towards the Atalanta Planitia lowlands. Despite its great length – longer than the Earth's Nile River – this remarkable channel has a uniform width of around 1.8 kilometres. Both of its ends are hidden by lava flows, so its true extent is not precisely known.

There's plenty of evidence that lava channels have produced erosion into pre-existing lava flows. In addition, meanders, braids and relict channels can be seen here and there, providing evidence that the channels formed over an extended period

Various volcanic features including lava flow fields, lava channels, anemone shields, calderas, in addition to a criss-crossing network of fissures can be seen in this image of Atla Regio (image approximately 300 km across). NASA.

Streamlined islands in a lava outflow channel in the Ammavaru area, Lada Terra. The tear-shaped islands point in the lava flow direction. NASA.

of time and resulted from more than one outpouring of lava. Some Venusian lava channels appear to run within a clearly defined broader band of terrain containing braids and relict channels, taking on a strong resemblance to a terrestrial river surrounded by oxbows and abandoned channel segments within a broader flood plain. Volcanic depressions are often found in association with sinuous rilles, forming a 'lake' from which many sinuous rilles emanate. Irregular in outline with

A sinuous lava channel some 200 km long and 2 km wide cuts through the terrain south of Atira Mons. NASA.

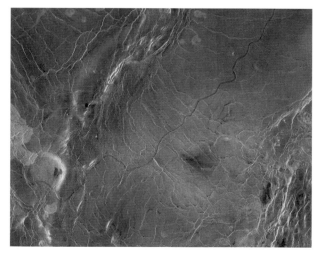

A 600 km long section of the Solar System's longest lava channel, Baltis Vallis. NASA.

rounded edges, these depressions in the surface are the product of lava pooling above a volcanic vent, with erosion and/or subsequent surface subsidence as activity dwindled.

Tectono-Volcanic Structures

In addition to clearly recognizable volcanic eruptive features, volcanism has combined with tectonic activity, crustal stresses and faulting over rising magma to produce complex tectono-volcanic features. Various types of these structures, differentiated by the nature of faulting, have been identified on Venus.

Calderas

Venus shows numerous instances where large intrusive magma chambers have risen into the lithosphere, deforming it into a dome, producing crustal tension and compression which has been translated into folds, crustal extension, concentric faults, fissures and rilles. Often this has been associated with a flurry of volcanic activity and dyke formation. In many cases the magma chamber has ceased to rise, cooled prematurely and withdrawn, producing a collapse in the magma chamber's roof and forming a large circular or elongated depression in the planet's surface known as a caldera. Often found in highland areas, Venusian calderas can easily be distinguished from impact craters owing to their lack of a blocky raised rim and broad terraced internal walls; they also lack inner ring features or a pronounced central elevation, and calderas are not surrounded by blankets of ejecta.

Calderas are unknown on the Moon and Mercury, but there are many terrestrial examples (although they are much smaller than those of Venus). The largest, best-preserved caldera on Earth is the Ngorongoro Crater in Tanzania, some 20 km across and 610 m deep, with steep inner walls that enclose an area of 260 square kilometres. Venusian calderas are similar in appearance but generally much larger than the Ngorongoro Crater.

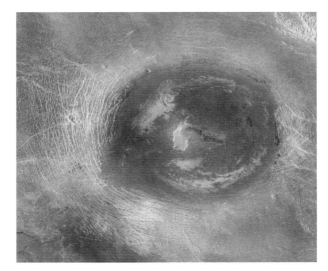

Sacajawea Patera (64N, 335E), an elliptical caldera measuring 175 × 260 km. Its floor lies some 2,000 m beneath its rim. NASA.

Coronae

A striking feature of the radar-imaged Venusian landscape, coronae are large calderas encircled by concentric faults and ridges, along with a radial pattern of faults and lava flows. Coronae are generally elliptical in outline and range between 100 and 1,000 km in diameter, although most of them average between 200 and 250 km across. Forming through upwelling plumes of magma which elevate and fault the surface, volcanic activity is relatively short-lived and too brief to completely bury the uplifted surface. Small volcanic features may make an appearance during the short period of activity; lava flows and pancake volcanoes are often seen in and around the coronae. Once the intrusion cools, the central elevation subsides, producing a caldera with steep inner walls and further fault features.

More than 200 Venusian coronae are known, and most of them occupy the higher parts of Venus' plains in areas that have experienced regional crustal compression. Many occur in chains or in small clusters, notably around the Beta Regio/Atla Regio/Themis Regio area, along the Parga Chasmata and near Hecate Chasma. However, the global distribution of these groups is uneven; there are fewer coronae groups to be found in the southern and eastern hemispheres between longitudes 0 and 180E.

Fotla Corona (59S, 164E), 150 km in diameter, displays a number of 'pancake' volcanoes. NASA.

Arachnoids

Taking their name from a spider or spider web-like appearance, arachnoids are considered to be a smaller-scale counterpart of coronae, sharing the same mode of origin. Arachnoids are circular structures that range between 50 and 250 km across. Like coronae, they display a ring of faults and ridges along with radial faults and ridges, but their rings generally completely enclose the radial features. It is possible that many of the radial features may represent dykes – narrow intrusions of magma along crustally weakened pathways. In any case, arachnoids display less evidence of volcanic activity and lava flows within them are rare. Of the 250 known arachnoids, around 60 of them lie within one of four large clusters. Somewhat confusingly on the maps, there's an overlap of official feature designations, and no separate designation for arachnoids. Most of the Venusian arachnoids come under the 'corona' label, some others are labeled 'paterae' while one is officially a Venusian mons.

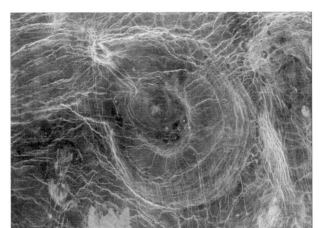

A typical arachnoid feature on Venus. NASA.

Novae

Also known as 'stellate fracture centres', novae consist of a pattern of faults and/or dykes arranged in a 'starburst' pattern arranged about a domoidal uplift. Of a similar size range to arachnoids, most novae are between 150 and 200 km in diameter. They display little evidence of volcanism. It is likely that these features, of which more than 50 are known, represent the first stages of arachnoid formation.

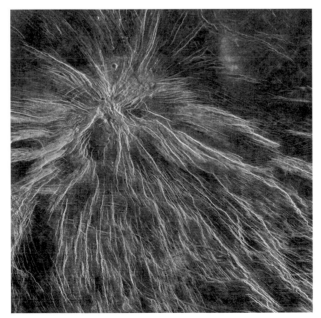

This nova lies in Themis Regio at 30S, 279E and measures 250 km across. NASA.

Pit Chains

These chains of small craters or pits are sometimes seen along grabens or in places of crustal extension. Superficially resembling secondary impact chain craters, pit chains are formed by surface collapse as a fault valley widens through crustal tension.

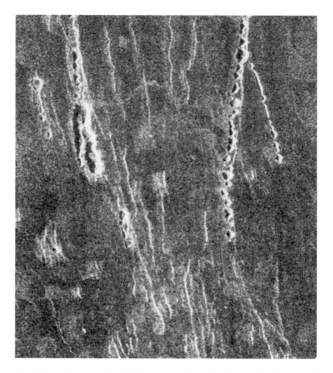

A typical series of small crater pit chains found along faults and graben on Venus. NASA.

Impact Features

Venus has such a geologically fresh surface – most of it younger than 500 million years old – that it is not possible to trace any really ancient impact features. Nothing of the great bombardment, so clearly visible in impact features on the Moon, Mercury (and to some extent, Mars) can be detected on Venus. Its youngest craters are younger than the vast majority of lunar craters visible through a small telescope, younger by half than the big, bright lunar rayed crater Copernicus.

The 1,000 or so Venusian impact craters – all of which were formed within the last 500 million years – are distributed in a random but uniform fashion over the planet's surface. There are only nine Venusian craters larger than 100 km in diameter; Mead (13N, 57E), the largest crater on Venus, measures 270 km across, around the same size as the 1.8 billion year old Sudbury impact crater in Ontario, Canada, the Earth's second largest impact feature.

Venus' largest impact craters

Crater	Centre	Diameter (km)
1. Mead	13N, 57E	270
2. Isabella	30S, 204E	175
3. Meitner	56S, 322E	149
4. Klenova	78N, 105E	141
5. Baker	63N, 40E	109
6. Stanton	23S, 199E	107
7. Cleopatra	66N, 7E	105
8. Rosa Bonheur	10N, 289E	104
9. Cochran	52N, 143E	100
10. Sayers	68S, 230E	98
11. Maria Celeste	23N, 140E	98
12. Potanina	32N, 53E	94
13. Greenaway	23N, 145E	93
14. Bonnevie	36S, 127E	92
15. Joliot-Curie	2S, 62E	91
16. Addams	56S, 99E	87
17. Sanger	34N 289E	84
18. Stowe	43S, 233E	80
19. Mona Lisa	26N, 25E	79
20. O'Keeffe	25N, 229E	77
21. Barsova	61N, 223E	76
22. Graham	6S, 6E	75
23. Wheatley	17N, 260E	75
24. Markham	4S, 156E	72
25. Boulanger	27S, 99E	72
26. Boleyn	24N, 220E	70
27. Henie	52S, 146E	70

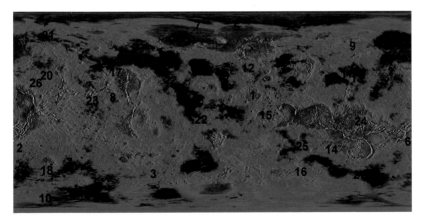

Locations of the largest of Venus' impact craters (key to numbering in the list above). NASA / Grego.

Venus has fewer small craters than any other planet in the Solar System; the smallest impact craters are all larger than about 1.5 km across, of the same sort of size as the Barringer 'Meteor' Crater in Arizona in the United States. This can be explained by the fact that Venus' substantially thick atmosphere prevents all but the largest meteoroids from impacting on the surface. It is estimated that any iron meteoroid smaller than around 30 m in diameter will burn up in the atmosphere and fragment or vaporize during its superheated descent.

The Impact Mechanism

Only sizeable meteoroids and small asteroids have been capable of penetrating Venus' atmosphere to strike the planet's surface. Those impactors big enough to make it to the ground and slice into the Venusian crust generate tremendous pressures and temperatures as their kinetic energy (the product of a meteoroid's mass and the square of its velocity) is converted into shock waves and heat which is imparted into the surrounding crust. The crust beneath the impactor is compressed and the surrounding material is pushed downwards and outwards. An ultra-hot bubble of expanding molten material with a temperature of several million degrees is formed as the impactor and the surrounding rocks are almost instantaneously vaporized. The edge of the crater is deformed and uplifted as a plume of excavated material, made up of vaporized rock and larger rock fragments, is blasted outwards from the impact site. As the crust decompresses, rebound effects produce a central uplift in larger craters, and a substantial layer of melted rock accumulates on the new crater's floor.

The excavated material is distributed around the crater in an ejecta blanket. The first material to be ejected comprises material that was close to the focus of impact near the surface, and this high velocity material is launched steeply above the surface

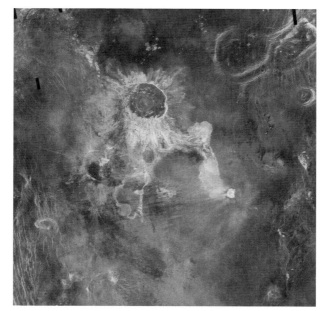

Venus' second largest impact crater, Isabella (175 km across) displays two extensive flow-lobes to its south and southeast. NASA.

to be deposited furthest from the crater. As the impact progresses, deeper material is excavated, but the overall energy of the impact dissipates. With progressively slower velocities the ejecta is distributed ever closer to the crater, and the deepest excavated bedrock may barely be lobbed over the crater's rim.

Ejecta patterns of Venusian craters are quite unlike the radial ejecta forms found on the Moon or on Mercury. Much Venusian ejecta is distributed in bright radar-reflective 'splotch' patterns around impact craters, many of which show a peculiar broad elongated tongue or extended finger of material. Unique to Venus, this type of feature is known as a flow lobe; they are the result of a mixture of hot gases, impact-melted ejecta and clastic material rapidly flowing downhill on one side of the crater immediately following the impact. They resemble pyroclastic flows arising from violent terrestrial eruptions, one of the best recent examples being the eruption of the Philippines' Mount Pinatubo in 1991.

Classification of Venusian Impact Craters

Crater Types

Seven morphological types of impact craters have been identified on Venus; these crater types and their size ranges closely match those found on the Earth.

Multi-ringed crater basins. Venus shows no impact features of the scale and age of the vast multi-ringed impact basins to be found on the Moon, Mercury and Mars, most of which formed during the great bombardment which ended 3.8 billion years ago. The largest example of a Venusian multi-ringed impact basin is Klenova, which has an ill-defined inner ring of 70 km diameter and two discontinuous

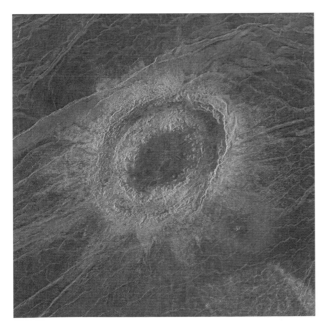

Multi-ringed impact basin Klenova (78N, 105E/141 km). NASA.

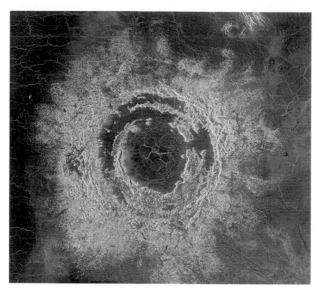

Double-ring crater Mona
Lisa (26N, 25E/79 km).
NASA.

concentric ring structures of 105 km and 141 km across. Smooth lava plains fill
its central portion and parts of the region between its outer rings. The outer ring
is made up of a series of small arcuate scarps similar to those seen in basins such
as the Moon's Orientale and Mercury's Caloris; radial ejecta and secondary impact
features surround the feature.

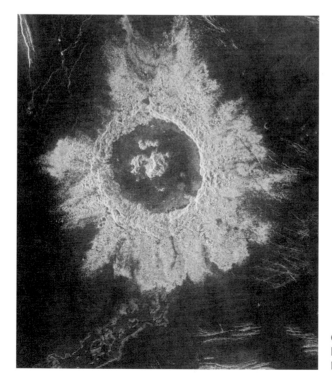

Central peak crater Dani-
lova (26S, 337E/49km).
NASA.

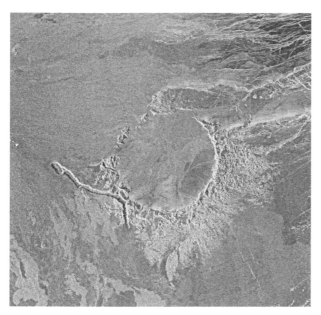

Alcott (60S, 355E/63 km), a crater with flooded, structureless floor. NASA.

Double-ring craters. Typically displaying a ring spacing ratio of 2:1, these impact features have a well-defined outer rim and an inner ring of some description. Most of Venus' craters larger than 40 km across fall into this category. Notable examples of this category are the craters Mona Lisa (79 km), whose floor is smooth and

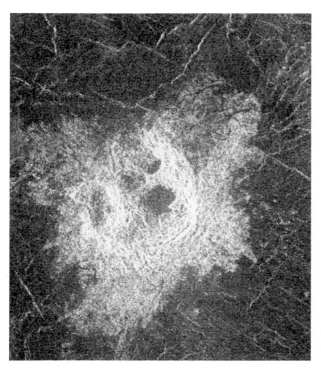

An unnamed irregular crater (21S, 335E/15 km). NASA.

An unnamed splotch (55N, 349E/30 km). NASA.

radar-dark, and Stanton (107 km), whose floor by contrast is relatively rough and radar-bright.

Central peak craters. Around one-third of Venus' craters have central elevations of some description, ranging from low mounds to prominent central mountains.

Craters with structureless floors. Many of the smaller (sub-15 km diameter) craters on Venus fall into this category. Many have rough, flattish radar-bright floors surrounded by terraced walls.

Irregular craters. Many of Venus' small impact craters have irregular outlines with rough, radar-bright floors. In many cases, these irregular features are actually conjoined clusters of craters formed by the impact of a body that broke up as it entered the Venusian atmosphere, each fragment impacting near the other almost simultaneously.

Splotches. A peculiar type of impact feature that planetary scientists have termed 'splotches' are seen in numerous locations on the Venusian plains. Splotches consist of an irregular-shaped area (often radar-dark), inside which can be found one or more craters or depressions. These features are produced by large meteoroids that have almost completely fragmented in a single, low-altitude air blast before hitting

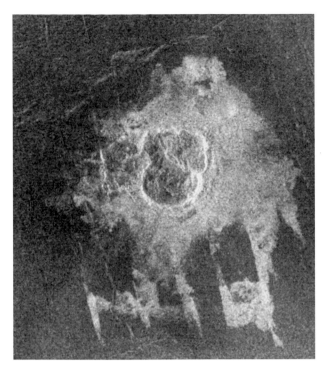

An unnamed multiple crater (16N, 352E/9 × 6 km). NASA.

the surface. Shock waves ripple across the surface, followed by a dense impacting cloud of small meteoroidal fragments which is deposited in across a portion of the surface, producing the irregular but well-defined splotch. Any surviving larger fragments impact to produce small craters within the splotch; these craters may be surrounded by a lighter halo of ejecta.

Multiple craters. Examples of this type of crater are to be found on every impacted body in the Solar System. Overlapping, conjoined, closely-spaced or chained craters are formed by the simultaneous impact of a number of objects (originally a single impactor that has broken up in space or during its descent towards the surface).

Post-impact Crater Modification

In addition, craters may be further classified by their different states of post-impact modification by volcanism, with lava flows embaying parts of the crater's floor or its outer ramparts, along with faulting and/or tectonic activity. According to the USGS Venus crater database, these basic states may be defined as:

Pristine. A crater and ejecta system that has been essentially unaltered since its formation.

Slightly embayed. A crater whose ejecta has been slightly embayed by lava.

Moderately embayed. A crater whose ejecta system has been moderately embayed by lava.

Pristine crater Stuart (31S, 20E/69 km). NASA.

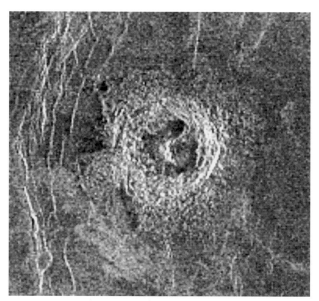

Ketzia (4N, 301E/15 km). NASA.

Bernadette (47S, 286E/13 km), a moderately embayed crater. NASA.

Heavily embayed. A crater whose rim, ejecta and (in many cases) floor have been heavily embayed by lava.

Heavily embayed and fractured. A crater heavily embayed by lava flows which also displays heavy fracturing.

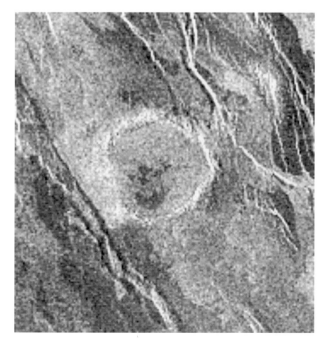

Raisa (28N, 280E/14 km), a heavily embayed crater. NASA.

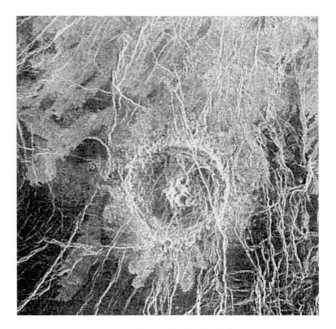

Baranamtarra (18N, 268E/26 km). NASA.

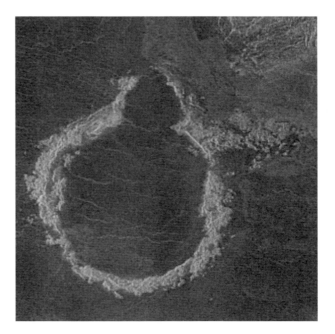

Double crater Heloise (40N, 52E/38 km), a very heavily embayed crater. NASA.

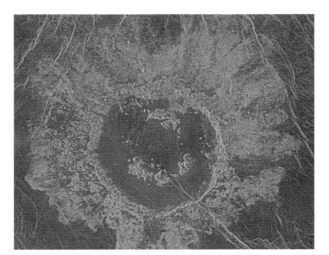

Wheatley (17N, 268E/75 km). NASA.

Very heavily embayed and fractured. A crater with a fractured rim whose ejecta and rim have been heavily embayed by lava flows.

Slightly fractured. A crater with fracturing affecting less than half of its floor, wall and rim.

Heavily fractured. A crater whose floor, wall, and rim are more than 50 percent affected by fractures.

Greatly disrupted. A crater that has been disrupted and degraded by tectonic forces following their formation. Very few are known, but they clearly show that some degree of crustal movement has taken place in Venus' relatively recent geological past.

Compressed fracture. A crater that has experienced slight fracturing through compressional faulting.

Ejecta mantled. A crater mantled by ejecta from younger impact feature nearby.

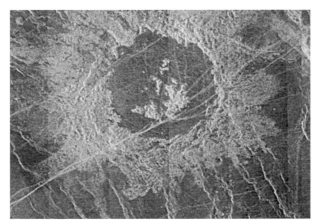

Tubman (24N, 25E/43 km), a heavily fractured crater. NASA.

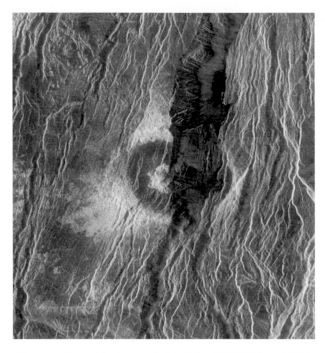

Tectonically modified crater Balch (30N, 283E/40 km). NASA.

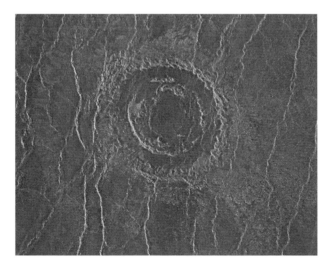

Barrymore (52S, 196E/57 km). NASA.

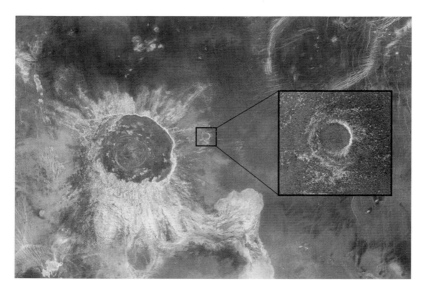

Alimat (30S, 206E/14 km), a crater mantled by ejecta from the younger and larger impact crater Isabella to its west. NASA.

Venus' Atmosphere

Venus' atmosphere is almost a thousand times thicker than that of the Earth. Not only does the planet have a denser and more substantial atmosphere than that of the Earth, it is of an entirely different composition. Carbon dioxide (CO_2) is the main component, making up some 96.4 percent of the Venusian atmosphere (compared

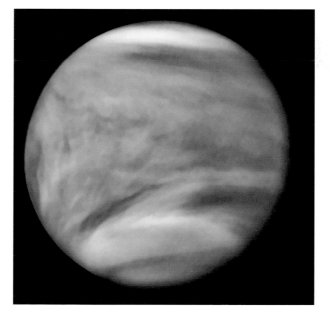

Venus' atmospheric banding is evident in this ultraviolet image taken by NASA's Pioneer Venus Orbiter in 1979. NASA.

to a terrestrial atmospheric presence of just 0.03 percent Co_2). 3.4 percent of Venus' atmosphere comprises nitrogen gas (compared to 78 percent N_2 in the Earth's atmosphere). The remaining constituents of Venus' atmosphere – sulphur dioxide, argon, water vapour, carbon monoxide, helium, neon, – are present in very small amounts.

Runaway Greenhouse Effect

About 35 percent of sunlight falling upon the Venusian cloud tops penetrates deeper into the atmosphere; most of it is absorbed by the clouds, but two percent of sunlight makes it to the ground and is absorbed by the planet's surface. As the surface heats up it emits mainly infrared radiation; since carbon dioxide and other atmospheric constituents are opaque to infrared radiation the heat is trapped. This process, known as the greenhouse effect (although the analogy with a glass garden greenhouse is not entirely accurate) has run away on Venus to its extreme limit, producing a surface temperature of around 500°C. There's little variation in surface temperatures between the planet's day and night sides, even though a point on the equator experiences 121 days of night and 121 days of light; the thick hot atmosphere evens temperatures out.

Such hellish surface conditions are unlikely to change naturally for several billion years – until the Sun's output of energy changes dramatically. However, for decades it has been speculated that by means of 'terraforming' the planet might in the far future be transformed into an inviting world capable of sustaining human life. This might be achieved within a few centuries by cooling the planet down and by biologically modifying the atmosphere.

How did Venus develop such lethal conditions? Early in the planet's history, Venus may have resembled the Earth, with planet-wide oceans of water. Being closer to the Sun, Venus' atmospheric temperature was greater than that of the Earth, enabling a greater amount of water vapour to be stored in the atmosphere. Water vapour, arising from the evaporation of Venus' oceans, is an effective greenhouse gas, causing the atmosphere to heat up yet further. A cycle of increasing evaporation from the oceans in tandem with increasing atmospheric temperature continued until all the water on the planet's surface evaporated and was absorbed by the atmosphere. It is thought that the entire process of oceanic evaporation on Venus might have taken place in as little as 600 million years. Most of the water in Venus' atmosphere was broken down by ultraviolet radiation in sunlight; the hydrogen atoms drifted off into space and the oxygen atoms were recycled to oxidize certain minerals on the surface.

Through the Atmosphere

With temperatures more than 200°C hotter than a standard kitchen oven at its highest setting and an atmospheric pressure of up to a hundred times that found at the Earth's surface, the lowest portion of Venus' atmosphere is an extremely inhospitable environment. Surface winds in this stifling, baking environment are sluggish, less than 7 km/hour. Temperature and pressure decreases with altitude. At a height of some 55–65 km above the planet's surface, the range of atmospheric temperature and pressure is the most Earth-like in the entire Solar System.

Sulphuric acid, produced as a result of the chemical combination between sulphur dioxide and the tiny amounts of water in the planet's atmosphere, forms a sulphuric haze at a altitude of between around 20–50 km above the surface. Above this is a far thicker layer of sulphuric acid clouds, the tops of which reach heights of around 65 km on the day side and 90 km on the night side, beyond which lies a more diffuse sulphuric haze.

Quantities of atmospheric sulphur dioxide have been found to vary by a factor of ten, indicating that there is an active input of great quantities of gases from a number of very large Venusian volcanoes. Lightning flashes detected over these areas may be due to the generation and discharge of huge amounts of static electricity among wind-borne particles of volcanic ash and dust.

Venus' clouds reflect some 60 percent of the sunlight that falls on them, giving the planet its high albedo and brilliant appearance; they also prevent the surface from ever being seen visually. At cloud top levels a high speed jet-stream blows from west to east at velocities of between 300 and 400 km/hour. Fastest at the equator and slower toward the poles, the high winds drive Venus' clouds around the planet in a period of around four days, frequently producing a 'V' type pattern in the clouds that can be observed visually. Observations have also revealed hurricane-like vortices in the cloud patterns over the planet's polar regions. In 2006 an unusual double vortex over Venus' south pole was observed at a height of 59 and 70 km. While vortices produced by the planet's jet-streams are predicted in models of the atmosphere, the dual nature of this particular vortex currently remains difficult to account for, suggesting that the planet's atmospheric dynamics are more complex than previously thought.

Magnetic Field

Even though Venus' internal structure is thought to closely resemble the Earth's interior, rather surprisingly the planet's intrinsic magnetic field has a maximum strength of a mere 1/100,000th that of the Earth. Two major factors might be responsible for the planet's lack of a strong internally-generated magnetic field – its slow rotation on its axis and the state of its core. The Earth's global bipole magnetic field is thought to be produced by a dynamo effect consequent upon convection currents within the ductile outer iron core (a conducting fluid) and the core's rotation. Venus may once have had an intrinsic magnetic field as strong as the Earth's early in its history – perhaps during the first billion years of its existence – although fossil magnetism cannot be seen because of the relatively young age of the planet's surface.

Venus' ionosphere does however interact with the solar wind to produce a weak magnetosphere, with a magnetic wake that trails behind the planet on the opposite side to the Sun. Venus' induced magnetosphere is nothing like as strong or as complex as the intrinsic magnetosphere of the Earth or Mercury; it doesn't possess a clearly-defined magnetotail, nor are there any radiation belts of trapped particles such as those found around the Earth. Instead, the profile of Venus' ionospheric magnetic boundaries changes with solar activity; the boundary can drop to an altitude as low as 250 km during periods of high solar activity.

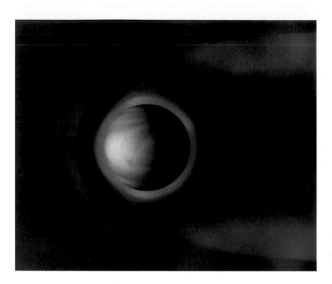

Venus' weak magneto-sphere is induced by solar activity.

Surface Conditions

Venus' average distance from the Sun is more than 50 million kilometres further than Mercury, and area for area Venus receives just 25 percent of the solar energy received by Mercury. Despite this, Venus' surface is generally somewhat hotter than the hottest subsolar point on Mercury. Temperatures on Venus can reach a searing maximum of 773 Kelvin – enough to melt tin, zinc and lead – several tens of degrees Kelvin hotter than that experienced at midday in Mercury's Caloris basin. The range of temperatures on the surface of Venus is moderated by its atmosphere, and as a result, temperature does not fall dramatically during the night – temperatures are sustained and maintained at a high level, day and night, because of the planet's dense atmosphere and a runaway greenhouse effect which traps solar energy.

If Venus had an Earth-like atmosphere then its surface would be around twice as brilliantly illuminated as the Earth. However, Venus' midday surface illumination is just 5,000 lux with a visibility of three kilometres – about the same as Moscow on an overcast mid-winter's day.

Venus' surface is far too hot to host water ice or liquid water; all of the planet's water takes the form of atmospheric water vapour. With a total mass of just 1/650th of the Earth's stocks of water, if Venus' atmospheric water were converted to a planet-wide ocean it would make up a liquid layer averaging just three metres deep.

The atmospheric pressure at Venus' surface is a staggering 90 kg per square centimetre – equivalent to the pressure experienced at a depth of 1 km beneath the ocean's surface. Winds blowing across the surface are sluggish, with average speeds of between 1 and 3.6 km/hour. However slow the winds, the atmosphere's high density enables soil particles to be slowly transported across the surface, forming depositional features like dunes in addition to producing erosion. Radar-bright wind streaks can be seen on the lee side of numerous features of high relief on Venusian plains.

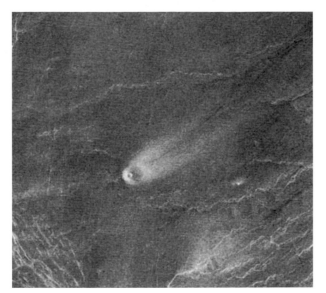

A 35 km long wind-blown streak of dust lies on the lee side of a small Venusian volcano. It superficially resembles an asymmetric ejecta deposit from an impact crater, like the Moon's Messier crater. NASA.

A Survey of Venus

It's not surprising that Venus takes its name from the ancient Roman goddess of love. With the unaided eye the planet frequently shines brilliantly in the evening or morning skies, appearing like a beautiful white lantern suspended in the air. It is fitting that the only planet named after a female deity has been given feminine nomenclature for its surface features – with the exception of Maxwell Montes, a giant mountain massif named in honour of the Victorian-era physicist James Clerk Maxwell.

Venusian Coordinates

Convention determines that longitude on a Solar System body should be measured from an arbitrary prime meridian, increasing in value in a direction opposite to that body's axial rotation. Since Venus rotates in an opposite direction to other planets, longitude is measured eastwards from the planet's prime meridian (0°E). The centre of a small crater, Ariadne in Sedna Planitia, is used to mark Venus' prime meridian.

Venusian Nomenclature

Venus displays a wide range of topographic features, most of which are the product of volcanism, localized crustal movement and/or tectonic activity. The planet has far fewer impact craters than Mercury or Mars. Most of the features listed below represent a classification based on topographic appearance, but most features in any group share the same mode of origin (as outlined above).

Arcus (arcus)	Arcuate feature(s).
Astrum (astra)	Radial feature(s).
Caldera	A rounded, steep-sided depression in the surface.
Catena (catenae)	Crater chain(s).
Chasma (chasmata)	An elongated steep-sided depression.
Colles	A knoll.
Corona (coronae)	Ovoid feature(s).
Crater	A circular cavity in the surface caused by impact.
Dorsum (dorsa)	Ridge(s).
Farra	A flat, pancake-like structure.
Fossa (fossae)	Elongated shallow depression(s).
Fluctus (fluctus)	Flow terrain.
Labyrinthus	A complex of intersecting valleys.
Linea (lineae)	An elongated marking (radar bright or dark).
Mons (montes)	A mountain (mountain range).
Patera (paterae)	Irregular or complex crater(s) with scalloped edges.
Planitia (planitiae)	Low plain(s).
Planum (plana)	Plateau(x) or high plain(s).
Regio (regions)	Large, clearly defined area(s) of distinct radar reflectivity.
Rupes (rupes)	Scarp(s).
Tessera (tesserae)	Polygonally fractured terrain with a mosaic-like appearance.
Terra (terrae)	Extensive landmass(es).
Tholus (tholi)	Isolated rounded hill(s) or mountain(s).
Undae	A system of dunes.
Vallis (valles)	A valley (valley system).

Some notable examples of Venus' topographic features

Type	Example	Size (km)	Centre
1. Chasmata	Parga Chasmata	11,000	20S, 255E
2. Colles	Akkruva Colles	1,059	46N, 116E
3. Corona	Artemis Corona	2,600	35S, 135E
4. Dorsa	Vedma Dorsa	3,345	42N, 159E
5. Farra	Seoritsu Farra	230	30S, 11E
6. Fluctus	Vut-Ami Fluctus	1,300	38S, 67E
7. Fossae	Karra-mahte Fossae	1,800	28N, 342E
8. Impact crater	Mead	270	13N, 57E
9. Labyrinthus	Radunitsa Labyrinthus	100	9S, 351E
10. Linea	Morrigan Linea	3,200	55S, 311E
11. Mons	Var Mons	1,000	1.2N, 316E
12. Multi-ringed impact basin	Klenova	141	78N, 105E
13. Patera	Sacajawea Patera	233	64N, 335E
14. Planitia	Guinevere Planitia	7,520	22N, 325E
15. Planum	Lakshmi Planum	2,345	69N, 339E
16. Regio	Eistla Regio	8,015	11N, 22E
17. Rupes	Vaidilute Rupes	2,000	44S, 22E
18. Terra	Aphrodite Terra	10,000	6S, 105E
19. Tesserae	Sudenitsa Tesserae	4,200	33N, 270E
20. Tholus	Toci Tholus	300	30N, 355E
21. Undae	Ningal Undae	225	9N, 61E
22. Vallis	Baltis Vallis	6,800	37N, 161E

A Venusian Triptych

The following topographic survey of the entire surface of Venus is divided into three regions of equal area which cover both northern and southern hemispheres in 120° wide longitudinal sections. These areas are the Ishtar-Alpha-Lada Region (300 to 60E), the Niobe-Aphrodite-Artemis Region (60 to 180E) and the Kawelu-Atla-Helen Region (180 to 300E).

Each of these three regions is surveyed in a general north to south, west to east trend, using the larger topographic features as the main points of reference in the text. Features inside or close to the main reference features are surveyed in a general clockwise trend from north, through west. In order to help the flow of the narrative, these rules are of a general nature, and there are a number of diversions and some overlaps with adjoining regions where necessary. Initial references to each particular feature are in **bold face** and in most cases are followed by the latitude and longitude (in parentheses) of the feature's central point, to the nearest degree; this information is often accompanied by the feature's diameter or main dimension(s).

Feature names, named feature coordinates and the dimensions of named features have been derived from the US Geological Survey's Astrogeology Research Program website Gazetteer of Planetary Nomenclature at http://planetarynames.wr.usgs.gov/index.html

Accompanying each quadrant line map is a tour map showing the general course of the survey of Venus' topographic features. Providing the text is read with occasional reference to these maps and the images that accompany it, the reader will be in little danger of losing themselves on this fascinating, almost Earth-sized world.

Region One: The Ishtar-Alpha-Lada Region (300 to 60E)

Centred on the Venusian prime meridian just east of the huge **Heng-o Corona** (2N, 355E/1,060 km), much of the central portion this area is covered by vast sweeping plains, notably **Sedna Planitia** (43N, 341E/3,570 km) in the north and **Lavinia Planitia** (47S, 348E/2,820 km) in the south. Near the equator lie patches of hilly, wrinkled terrain underscoring **Eistla Regio** (11N, 22E/8,015 km); the isolated rough highland patch called **Alpha Regio** (26S, 0E/1,897 km) rises abruptly from the plains in the central southern portion of the region. Counterbalancing these features are two large continental regions – the imposing highland plateau of **Ishtar Terra** (70N, 28E/5610 km) in the far north and the larger but more vertically-restrained **Lada Terra** (63S, 20E/8,615 km) in the far south.

Louhi Planitia (81N, 121E/2,440 km) occupies much of the far north of the region beyond the northeastern borders of Ishtar Terra. Several ridges cross its surface, the largest of which, **Tezan Dorsa** (81N, 47E) stretches for more than 1,000 km across the plain. To its west, **Snegurochka Planitia** (87N, 328E/2,775 km) forms a pronounced indent in the northern edge of Ishtar Terra.

Ishtar Terra, a vast continental highland region, sprawls across the northern part of the region for 5,610 km between longitudes 300 and 80E. It has an area of around 8 million square kilometres, equivalent to the area of Australia; unlike

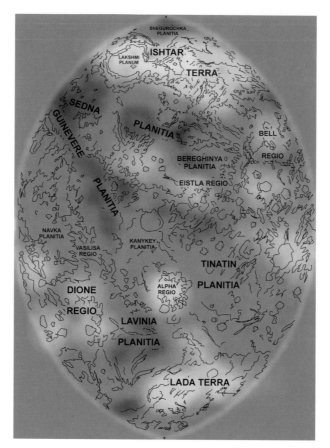

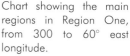

Chart showing the main regions in Region One, from 300 to 60° east longitude.

Australia, however, Ishtar Terra is topographically very varied, comprising three distinct sections. Its western component is the large mountain-encircled plateau **Lakshmi Planum** (69N, 339E/2345 km); the central component is presided over by the giant mountain massif **Maxwell Montes** (65N, 3E/797 km) which straddles the Venusian prime meridian; the eastern section is made up of the extensive hilly tesserated terrain of **Fortuna Tessera** (70N, 45E/2,801 km).

Lakshmi Planum, a vast plateau bordered by high mountains, bears a strong though superficial resemblance to the Earth's Tibetan Plateau. It is oval in shape, widest from west to east, and reaches a height of between 2,500 and 4,000 m. Two large volcanic domes with calderas rise from the plateau – **Colette Patera** (66N, 323E/149 km) to the west and the coronal **Sacajawea Patera** (64N, 335E/233 km) to the east. Colette Patera, the source of numerous large radial radar-bright lava flows, has a flat-based elongated caldera measuring 90 × 50 km, surrounded by concentric fractures. Sacajawea Patera, the largest feature of its type on Venus, is a corona-type feature with an elongated caldera measuring 120 × 215 km which is surrounded by a system of faults, graben and scarps, particularly densely packed in the west. On Sacajawea Patera's eastern flanks can be found a set of linear ridges which were possibly formed by crustal rifting and volcanism.

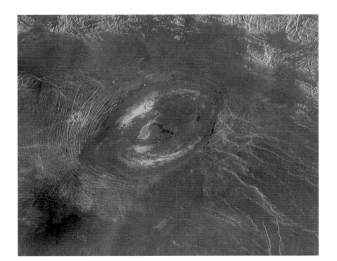

Sacajawea Planitia.
NASA/Magellan.

Forming a 579 km long border to the north of Lakshmi Planum the mountain range of **Freyja Montes** (74N, 334E) rises around 1,000 m above the plateau to its south and some 3,000 m higher than the hilly 'continental shelf' of **Itzpapalotl Tessera** (76N, 318E/380 km) to its north. East of Lakshmi Planum, Maxwell Montes dominates the central portion of Ishtar Terra. Seen from above, this vast mountain range is shaped something like the head of a hummingbird with its narrow beak tapering off to the west towards Sacajawea Patera. The range contains the highest peaks on Venus, some of which rise more than 10,000 m above the lowlands of Sedna Planitia to its southwest. Crustal compression is thought to have produced Maxwell, in the same manner as many of Earth's mountain ranges. The range is made up of a series of parallel ridges between 2 and 7 km apart. Its western flanks are very steeply inclined to Lakshmi Planum, while to the east there are more gentle gradients as it descends to the level of Fortuna Tessera. The crater **Cleopatra** (66N, 7E/105 km) forms a prominent 'eye' to the 'hummingbird'. Its main rim is polygonal in outline, with a radar dark floor; offset to the northwest of centre inside the main ring is sunk a smoother, darker-floored inner ring some 40 km in diameter. Of considerable interest is a breach in Cleopatra's northeastern wall through which lava has flowed into **Anuket Vallis** (67N, 8E/350 km) at the edge of Fortuna Tessera.

Danu Montes (59N, 334E/808 km) curve along the southern edge of Lakshmi Planum, their foothills merging into **Clotho Tessera** (56N, 335E/289 km). Further west, the range blends into the huge curving step-like scarp **Vesta Rupes** (58N, 324E/788 km) which clearly defines the plateau's southwestern edge. To the northwest, the striated appearance of the ridges in **Akna Montes** (69N, 318E/830 km) provides evidence that the range was formed as a result of crustal compression.

South of Fortuna Tessera can be found **Haumea Corona** (54N, 22E/375 km), one of the largest named volcanic radial fracture features on Venus; there is only a hint of concentric pattern of fracturing, indicating that this is an example of an early-stage corona formation. Towards Haumea runs the northeast-southwest

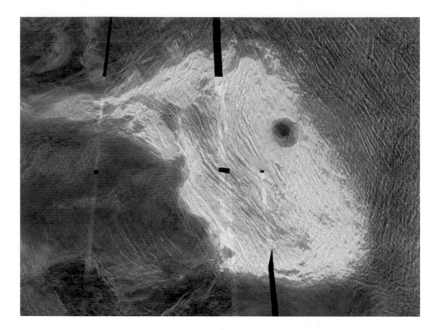

Maxwell Montes. NASA/Magellan.

oriented **Sigrun Fossae** (51N, 18E/970 km). Like many features on Venus, there is ample evidence that these features were produced in more than one phase of crustal deformation, along with **Aušra Dorsa** (49N, 25E/859 km) to their east.

A broad upland extension of Ishtar, containing **Laima Tessera** (55N, 49E/971 km), projects southwards from Fortuna Tessera. Laima's formation is possibly

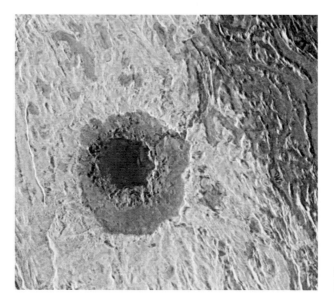

Crater Cleopatra. NASA/Magellan.

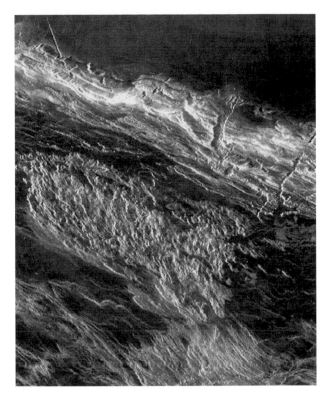

Vesta Rupes. NASA/Magellan.

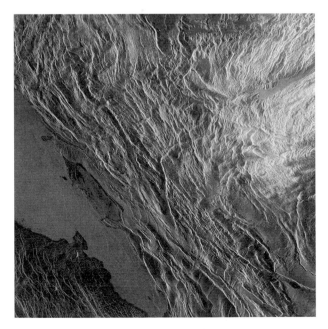

Akna Montes. NASA/Magellan.

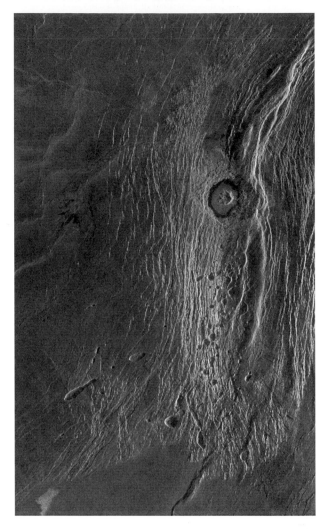

Close-up view of Akna Montes showing the 18 km crater Wanda, onto whose floor has collapsed a quantity of ridge material. NASA/Magellan.

linked to that of **Leda Planitia** (44N, 65E/2,890 km), which likely developed through an upwelling mantle plume, with downcurrents at its periphery casing crustal buckling, ridge formation and fracturing.

The fairly bland, smooth lowland plains of Sedna Planitia sweep across the northwestern part of the region. Of the few features within the plain itself, most noteworthy are **Beiwe Corona** (53N, 307E/600 km), **Xilonen Corona** (51N, 321E/300 km), **Bethune Patera** (47N, 321E/94 km) and **Sachs Patera** (49N, 334E/65 km). The two most lowlying portions of Sedna Planitia are separated by a slight north-south rise, atop of which runs **Zorile Dorsa** (40N, 338E/1,041 km), a compressional feature which extends south into the most westerly parts of Eistla Regio. A hilly area marks the eastern border of Sedna Planitia, inside which are to be found a cluster of volcanic features, including **Ashnan Corona** (50N, 357E/300 km), **Ba'het Corona** (48N, 0E/145 km) and **Onatah Corona** (49N, 6E/298 km).

Sedna's southern border is marked by the northwestern extension of Eistla Regio. The area contains a number of coronae, including **Renenti Corona**

(33N, 326E/200 km), nova **Mesca Corona** (27N, 343E/190 km), **Purandhi Corona** (26N, 344E/170 km), **Tutelina Corona** (29N, 348E/180 km), **Nissaba Corona** (26N, 356E/300 km) and **Idem-Kuva Corona** (25N, 358E/230 km). Two sizeable volcanoes lie to their south – **Sif Mons** (22N, 352E/300 km) and **Gula Mons** (22N, 359E/276 km). Sif rises to a height of 2,000 m and is topped by a 50 × 40 km caldera, the rim of which contains smaller nested calderas. Gula, 3,000 m high, has a double caldera measuring 40 × 30 km. The steep flanks of both volcanoes are covered with radar-dark lava flows.

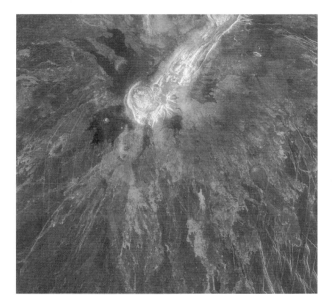

Gula Mons. NASA/Magellan.

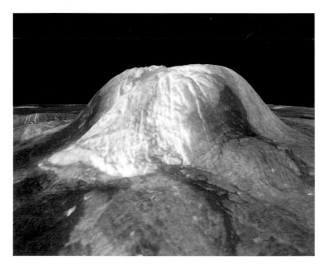

Simulated view of Gula Mons from the southwest. NASA/Magellan.

Guinevere Planitia (22N, 325E/7,520 km) sweeps in from beyond the western reaches of Sedna Planitia to touch the equator. With a hillier nature than its neighbour, Guinevere blends into **Undine Planitia** (13N, 303E/2,800 km) to its west, where there are numerous coronae, including the arachnoid **Madderakka Corona** (9N, 316E/220 km). **Var Mons** (1N, 316E/1,000 km), Venus' widest volcanic mountain, has three main cones of 1,500, 700 and 1,700 m in height; four flow fields surround them, the youngest of which emanes from the central cone. Adjoining Var to the northwest is another large volcanic mountain, **Atanua Mons** (10N, 309E/1,000 km), 1,600 m high and surrounded by a number of substantial lava flows.

Eastern Guinevere contains the fractured ring of **Benten Corona** (16N, 340E/310 km) and the vast 1,060 diameter Heng-o Corona, Venus' second largest corona. Inside Heng-o's faulted ring there are a number of small impact craters and an intricate delta-shaped system of two fracture sets, one trending to the north and the other trending to the northwest.

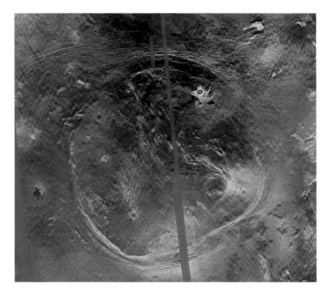

Heng-o Corona. NASA/ Magellan.

Beyond Sedna, continuing eastwards from the meridian, lies **Bereghinya Planitia** (29N, 24E/3,900 km), which is bordered in the east by tesserated terrain at the margins of **Bell Regio** (33N, 51E/1,778 km) and in the south by Eistla Regio. **Beyla Corona** (27N, 16E/400 km) occupies the smooth centre of Bereghinya, while its hiller eastern half is indented by the spectacular impact crater **Mona Lisa** (26N, 25E/79 km). Tesserated and ridged terrain borders Bereghinya to the north and east, with a series of northwest-southeast oriented features such as **Kruchina Tesserae** (36N, 27E/1,000 km) and **Hera Dorsa** (36N, 30E/813 km). These interest with **Bezlea Dorsa** (30N, 37E/807 km) and the hefty **Metelitsa Dorsa** (16N, 31E/1,300 km), a ridge and valley system on the western edge of Bell Regio.

Metelitsa Dorsa. NASA/
Magellan.

The eastern part of Eistla Regio displays considerable variety. **Guor Linea** (20N, 3E/600 km) a rift valley system, extends eastwards from the flanks of Gula Mons. A cluster of pancake domes nearby, **Carmenta Farra** (12N, 8E), measures 180 km across; the largest of these is 65 km in diameter and 1,000 m high. Several coronae and volcanic mountains lie further east, notable among them the 2,000 m high **Kali Mons** (9N, 29E/325 km) and **Dzalarhons Mons** (1N, 34E/120 km), to the east of which is the volcanic flow field of **Nekhebet Fluctus** (0N, 35E/400 km).

Bell Regio consists of two mountainous volcanic highlands separated by smooth low plains. Bell is unlike many among other Venusian volcanic uplifts because it does not display prominent rifting due to crustal deformation and faulting. In the northern part of Bell, a substantial volcanic rise features several large paterae and coronae, the largest being **Nefertiti Corona** (36N, 48E/371 km). Here the large volcano **Tepev Mons** (29N, 44E/301 km) rises to some 5,000 m above its surroundings. Like nearby **Nyx Mons** (30N, 49E/875 km), Tepev is closely encircled by a 'moat' several hundred metres deep, produced by crustal deformation due to the sheer weight of the volcanic mass. Southern Bell features a collection of large nova-type coronae, including the novae **Didilia Corona** (19N, 38E/320 km), **Pavlova Corona** (14N, 39E/370 km) and **Isong Corona** (12N, 49E/540 km) and

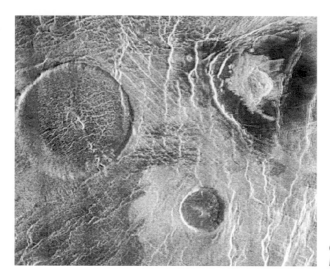

Carmenta Farra. NASA/ Magellan.

Calakomana Corona (7N, 44E/575 km). To their east lies **Mead** (13N, 57E/270 km), Venus' largest impact crater, a multi-ringed basin. Mead probably began as a crater of the dimensions of its central ring, with a rim some 200 km across; subsequent landslides around an outer fault ring have enlarged the cavity, creating a sharply-defined inner wall, a scarp 1,000 m high.

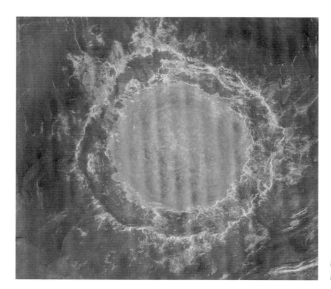

Mead crater. NASA/ Magellan.

South of Eistla Regio a line of evenly-spaced major coronae underscores the equator for 3,500 km, namely **Kuan-Yin Corona** (4S, 10E/310 km), **Thouris Corona**

(7S, 13E/190 km), **Cybele Corona** (8S, 21E/500 km), **Thermuthis Corona** (8S, 33E/330 km) and the conjoined corona **Mukylchin Corona** (13S, 46E/525 km) and the nova **Nabuzana Corona** (9S, 47E/525 km).

Between the longitudes of 300 and 60E, Venus' southern equatorial regions are made up of a patchwork of planitiae, from **Navka Planitia** (8S, 318E/2,100 km), through **Kanykey Planitia** (10S, 350E/2,100 km) and **Tinatin Planitia** (15S, 15E/2,700 km) to **Tahmina Planitia** (23S, 80E/3,000 km). These features contain the odd small volcano and corona, along with numerous low wrinkle ridges. **Bathsheba** (15S, 50E/32 km) is a prominent radar feature of northern Tahmina, an impact crater with a bright mantle of ejecta flows amid an extensive coma of westward-swept ejecta.

Dione Regio (32S, 328E/2,300 km) stretches across the southwest of the region; several volcanoes rise above its undulating surface. **Ushas Mons** (24S, 325E/413 km) in the north is 2,000 m high, flanked by radial lava flows and has a radar dark caldera. To the south of Dione, a cluster of volcanoes includes **Tefnut Mons** (39S, 304E/182 km), **Nepthys Mons** (33S, 318E/350 km), **Hathor Mons** (39S, 325E/333 km) and **Innini Mons** (35S, 329E/339 km).

Ushas Mons. NASA/ Magellan.

Lavinia Planitia, a vast plain to the southeast of Dione Regio, is crossed by several large lineae, including **Morrigan Linea** (55S, 311E/3,200), **Hippolyta Linea** (42S, 345E/1,500 km), **Antiope Linea** (40S, 350E/850 km), **Molpadia Linea**

(48S, 355E/1,600 km), **Penardun Linea** (54S, 344E/975 km) and **Kalaipahoa Linea** (61S, 337E/2,400 km), with some clear examples of strike-slip faulting through horizontal movement of the crust. **Mylitta Fluctis** (56S, 354E/1,250 km) on the southeastern border of Lavinia is one of Venus' largest lava flow fields.

Alpha Regio, a compact, well-defined mountain plateau straddles the Venusian prime meridian. Unlike Venus' major volcanic rises, plateau highlands such as Alpha Regio contain relatively few volcanic features such as domes, shields and lava flows; **Eve Corona** (32S, 360E/330 km) in the southwest is the only major volcanic feature associated with Alpha Regio. Instead, Alpha Regio comprises a roughly quadrilateral bumpy plateau of tesserated terrain of some 3.25 million square kilometres – the size of India – some 1,000 to 2,000 m above the surrounding smooth lava plains. It is thought that tesserated plateau highlands such as Alpha Regio may have formed when the crust was crumpled together, thickened and deformed during a process of cold spot mantle downwelling.

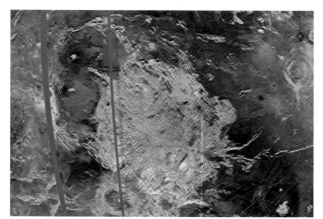

Alpha Regio. NASA/
Magellan.

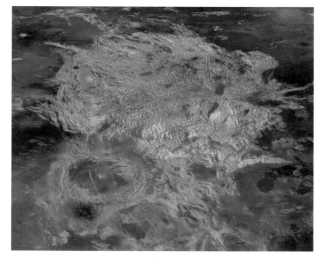

Simulated view of Alpha
Regio from the south.
NASA/Magellan.

A series of coronae south of Alpha Regio, including the nova **Carpo Corona** (38S, 3E/215 km), **Tamfana Corona** (36S, 6E/400 km) and **Seia Corona** (3S, 153E/225 km), plus the large valley **Hanghepiwi Chasma** (49S, 18E/1,100 km) extend towards the far southern continent of Lada Terra. Hanghepiwi Chasma is a rift valley between 500 and 1,000 m deep, filled with lavas from **Astkhik Planum** (45S, 20E/2,000 km) to the east. **Vaidilute Rupes** (44S, 22E/2,000 km), a large fault scarp, stretches from the north of Seia Corona around the northern and eastern edges of Astkhik Planum. Astkhik Planum itself indents the northwestern edge of Lada Terra, and to its east lies **Fonueha Planitia** (44S, 48E/3,000 km), an undulating plain whose northern border rises to encompass numerous coronae and the fault valley system of **Artio Chasma** (36S, 39E/450 km).

Lada Terra is an uplands region some 8,600 km across and averaging 2 km above the mean planetary radius, featuring numerous coronae, faults, graben and fluctus. The continent's highest areas are to be found in a large domed lobe to the southwest on the 0° meridian, dominated by **Quetzalpetlatl Corona** (68S, 357E/780 km) and **Boala Corona** (70S, 359E/220 km), parts of which reach heights of 4,000 metres. **Eithinoha Corona** (57S, 8E/500 km) and **Otygen Corona** (57S, 31E/400 km) occupy the low plains to the north of this rise. Two smooth plains – **Aibarchin Planitia** (73S, 25E/1,200 km) and **Mugazo Planitia** (69S, 60E/1,500 km) – separated by a volcanic rise containing **Okhin-Tengri Corona** (71S, 40E/400 km) radiate from Lada towards the south polar region. From **Ekhe-Burkhan Corona** (50S, 40E/600 km) the eastern part of Lada Terra rises towards the large rift valley system of **Xaratanga Chasma** (54S, 70E/1,300 km – see description in the Niobe-Aphrodite-Artemis Region below).

Quetzalpetlatl Corona. NASA/Magellan.

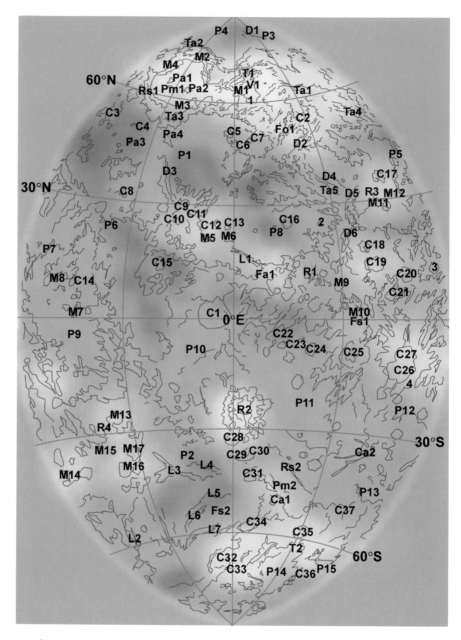

Map of Region One (300-60E), the Ishtar-Alpha-Lada Region, showing the locations of features described in the text.

Key to Features on Chart:

Pa Patera; Rs Rupes; Ta Tessera; Ca Chasma; Fo Fossa; Fs Fluctus; Fa Farra; Co Colles; M Mons; Pm Planum; P Planitia; C Corona; D Dorsa; T Terra; V Vallis; R Regio; C1. Heng-o Corona; C2. Haumea Corona; C3. Beiwe Corona; C4. Xilonen Corona; C5. Ashnan Corona; C6. Ba'het Corona; C7. Onatah Corona; C8. Renenti Corona; C9. Mesca Corona; C10. Purandhi Corona; C11. Tutelina Corona; C12. Nissaba Corona; C13. Idem-Kuva Corona; C14. Madderakka Corona; C15. Benten Corona; C16. Beyla Corona; C17. Nefertiti Corona;

Region Two: The Niobe-Aphrodite-Artemis Region (60 to 180E)

At first glance, this region of Venus might be summarized as two broad plains regions separated by the vast, sprawling sub-equatorial continent of **Aphrodite Terra** (6S, 105E/10,000 km). To the north lies a patchwork of planitiae separated by a collection of tesserae, dorsa and coronae; particularly noteworthy is **Tellus Tessera** (43N, 77E/2,329 km) an isolated raft of tesserated uplands of similar size and appearance to Alpha Regio (see Region One, above). A collection of planitiae also vie for position in the south, and in general these are smoother and less wrinkly than their northern counterparts. Several major chasmata wind their way across the southern sub-equatorial region, including **Juno Chasma** (31S, 111E/915 km), **Quilla Chasma** (24S, 127E/973 km), **Diana Chasma** (15S, 155E/938 km) and the magnificent **Artemis Chasma** (41S, 139E/3,087 km) which curves around the eastern edge of the vast **Artemis Corona** (35S, 135E/2,600 km).

The lowlands of **Louhi Planitia** (81N, 121E/2,440 km) stretch across the north polar region. Louhi's southwestern border comprises the tesserated terrain of eastern Ishtar Terra, while its southern edge is defined by the hills of **Tethus Regio** (66N, 120E/2,000 km). Tethus' eastern heights are dominated by **Nightingale Corona** (64N, 130E/471 km) and **Earhart Corona** (70N, 136E/414 km). Further east the terrain descends to the deep smooth lowlands of **Atalanta Planitia** (46N, 166E/2,050 km), its deepest area (located at 64N, 165E) comprising a 450 km diameter basin some 3,000 m beneath the mean surface level. Atalanta Planitia is likely to be a place of mantle downwelling, producing compressional ridge belts to its north and east.

C18. Didilia Corona; C19. Pavlova Corona; C20. Isong Corona; C21. Calakomana Corona; C22. Kuan-Yin Corona; C23. Thouris Corona; C24. Cybele Corona; C25. Thermuthis Corona; C26. Mukylchin Corona; C27. Nabuzana Corona; C28. Eve Corona; C29. Carpo Corona; C30. Tamfana Corona; C31. Seia Corona; C32. Quetzalpetlatl Corona; C33. Boala Corona; C34. Eithinoha Corona; C35. Otygen Corona; C36. Okhin-Tengri Corona; C37. Ekhe-Burkhan Corona; P1. Sedna Planitia; P2. Lavinia Planitia; P3. Louhi Planitia; P4. Snegurochka Planitia; P5. Leda Planitia; P6. Guinevere Planitia; P7. Undine Planitia; P8. Bereghinya Planitia; P9. Navka Planitia; P10. Kanykey Planitia; P11. Tinatin Planitia; P12. Tahmina Planitia; P13. Fonueha Planitia; P14. Aibarchin Planitia; P15. Mugazo Planitia; R1. Eistla Regio; R2. Alpha Regio; R3. Bell Regio; R4. Dione Regio; L1. Guor Linea; L2. Morrigan Linea; L3. Hippolyta Linea; L4. Antiope Linea; L5. Molpadia Linea; L6. Penardun Linea; L7. Kalaipahoa Linea; Fa1. Carmenta Farra; Fs1. Nekhebet Fluctus; Fs2. Mylitta Fluctis; T1. Ishtar Terra; T2. Lada Terra; D1. Tezan Dorsa; D2. Aušrā Dorsa; D3. Zorile Dorsa; D4. Hera Dorsa; D5. Bezlea Dorsa; D6. Metelitsa Dorsa; V1. Anuket Vallis; Pm1. Lakshmi Planum; Pm2. Astkhik Planum; M1. Maxwell Montes; M2. Freyja Montes; M3. Danu Montes; M4. Akna Montes; M5. Sif Mons; M6. Gula Mons; M7. Var Mons; M8. Atanua Mons; M9. Kali Mons; M10. Dzalarhons Mons; M11. Tepev Mons; M12. Nyx Mons; M13. Ushas Mons; M14. Tefnut Mons; M15. Nepthys Mons; M16. Hathor Mons; M17. Innini Mons; Ta1. Fortuna Tessera; Ta2. Itzpapalotl Tessera; Ta3. Clotho Tessera; Ta4. Laima Tessera; Ta5. Kruchina Tesserae; Pa1. Colette Patera; Pa2. Sacajawea Patera; Pa3. Bethune Patera; Pa4. Sachs PateraCa1. Hanghepiwi Chasma; Ca2. Artio Chasma; Rs1. Vesta Rupes; Rs2. Vaidilute RupesFo1. Sigrun Fossae; Craters: 1.Cleopatra; 2. Mona Lisa; 3. Mead; 4. Bathsheba

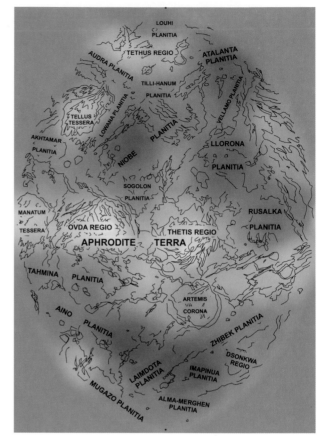

Chart showing the main regions in Region Two, from 60 to 180° east longitude.

Audra Planitia (60N, 92E/1,860 km) and **Tilli-Hanum Planitia** (54N, 120E/2,300 km) make up the hilly terrain around the south of Tethus Regio; from the latter rises **Fakahotu Corona** (59N, 106E/290 km), a heart-shaped set of fractures surrounded by a broader set of radial flows. Several patches of tesserated terrain border these plains, from **Dekla Tessera** (57N, 72E/1,363 km) in the west to **Ananke Tessera** (53N, 137E/1,060 km) in the west. Ananke displays clear evidence of crustal rifting, since its northern and southern sections, separated by an irregular gap of some 100 km in breadth, would fit comfortably together. The double-ringed crater **Cochran** (52N, 143E/100 km) lies on Ananke's eastern border.

Tellus Tessera is a polygonal plateau with an area of more than 4 million square kilometres, equivalent to that of the Sahara Desert. Like most other areas of tesserated terrain, Tellus represents an area of mantle downwelling, crustal compression and thrust faulting, followed by crustal relaxation and embayment by volcanic lavas. Two paterae, **Apgar Patera** (43N, 84E/126 km) and **Eliot Patera** (39N, 79E/116 km) can be found in the north of Tellus. Yet further north, a lowland lava plain, a flooded rift zone, occupies the space between Tellus and the smaller fractured plateau of **Meni Tessera** (48N, 78E/454 km). Beyond Tellus' northeastern margin is **Medeina Chasma** (46N, 89E), a straight valley for much of its 600 km length. Tellus is highest along its eastern margins, where a series of mountainous folds rise to 3,000 m above the mean surface level. Southern parts of Tellus also rise

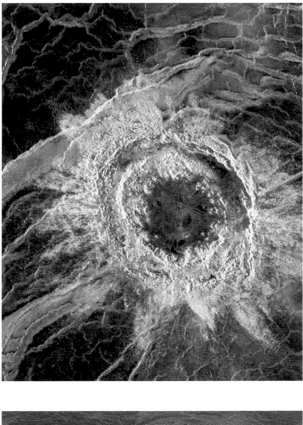

La Fayette (70N, 108E) a 40 km diameter impact crater in Tethus Regio. NASA/Magellan.

Part of the rifted Ananke Tessera. NASA/Magellan.

to heights of more than 2,000 m, gradually decreasing in height across a jumbled terrain towards the plains of **Lowana Planitia** (43N, 98E/2,700 km) which skirts the eastern and southern edges of the plateau. The valley **Kottravey Chasma** (31N, 78E) curves through Tellus' southwestern heights for some 744 km. Leda Planitia and **Akhtamar Planitia** (27N, 65E/2,700 km), separated by **Mardezh-Ava Dorsa** (32N, 69E/906 km), slope away to the west of Tellus Tessera.

   Niobe Planitia (21N, 112E/5,008 km), one of Venus' most extensive plains, straddles the mid-northern latitudes. Numerous northwest-southeast trending features cross Niobe, from **Akkruva Colles** (46N, 116E/1,059 km) in the north

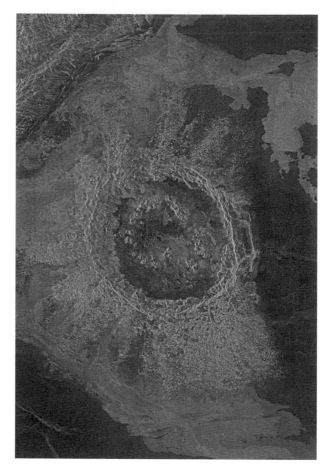

Cochran, a sizeable impact crater with a double ring. NASA.

to **Uni Dorsa** (34N, 114E/800 km), a set of ridges that connects **Kutue Tessera** (40N, 109E/653 km) to the northern tip of **Gegute Tessera** (17N, 121E/1,600 km). A cluster of small coronae occupies the southern reaches of Niobe and **Sogolon Planitia** (8N, 107E/1,600 km), including **Dhisana Corona** (15N, 112E/100 km), **Allatu Corona** (16N, 114E/125 km), **Bhumiya Corona** (15N, 118E/100 km) and **Omeciuatl Corona** (17N, 119E/175 km).

East of Gegute Tessera, both **Llorona Planitia** (18N, 145E/2,600 km) and the adjoining **Vellamo Planitia** (45N, 149E/2,155 km) to its north are wrinkled by a number of north-south trending ridges, from **Likho Tesserae** (40N, 134E/1,200 km), **Nephele Dorsa** (40N, 139E/1,937 km), **Frigg Dorsa** (51N, 151E/896 km) and **Vedma Dorsa** (42N, 159E/3,345 km), the latter being Venus' most extensive named system of wrinkle ridges. These dorsa extend across northern Llorona to the vicinity of several large coronae, namely **Boann Corona** (27N, 137E/300 km), **Cauteovan Corona** (32N, 143E/553 km) and **Ved-Ava Corona** (33N, 143E/200 km).

A cluster of impact craters **Maria Celeste** (23N, 140E/98 km), **Greenaway** (23N, 145E/93 km), **Callirhoe** (21N, 141E/34 km), **Bourke-White** (21N, 148E/34 km) lie in central Llorona in one of the least radar-reflective areas on the planet. To their south are several coronae, including **Abundia Corona** (19N, 125E/250 km) and **Kubebe**

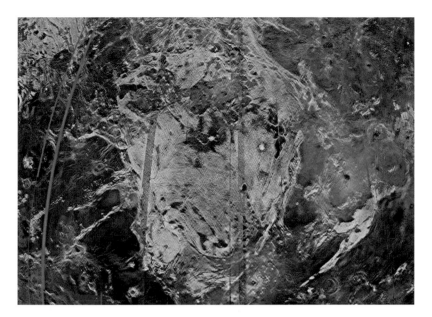

Tellus Tessera. NASA/Magellan.

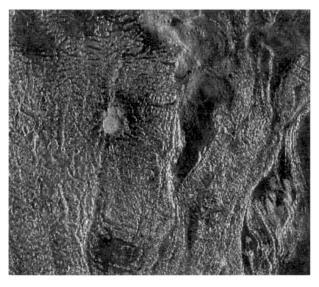

Part of northeastern Tellus Tessera. NASA/Magellan.

Corona (16N, 133E/125 km), while eastern Llorona hosts the impressive arachnoid **Ituana Corona** (20N, 154E/220 km) and its associated lava flow **Praurime Fluctus** (16N, 154E/750 km) which sprawls southwards into the northwestern plains of **Rusalka Planitia** (10N, 170E/3,655 km).

Rusalka Planitia is a lowlying plain bordered in the north by **Nemesis Tesserae** (40N, 181E/355 km) and **Athena Tessera** (35N, 175E/1,800 km). Rusalka has a southern central uplift dominated by **Lamashtu Mons** (3N, 173E/260 km), **Hannahannas Corona** (0N, 171E/200 km), **Nirmali Corona** (6S, 172E/60 km) and the arachnoids **Saunau Corona** (1S, 173E/220 × 160 km) and **Eigin Corona** (5S, 175E/200 km). Two

Gegute Tessera. NASA/Magellan.

extensive lava flows, **Argimpasa Fluctus** (0N, 176E/950 km) and **Dotetem Fluctus** (6S, 178E/530 km) spread down to the plains to the north and south of Eigin Corona. Large tracts of eastern Rusalka Planitia are crossed by many long north-south oriented dorsa, among them **Zaryanitsa Dorsa** (0N, 170E/1,100 km), **Yalyane Dorsa** (7N, 177E/1,200 km) and **Poludnitsa Dorsa** (5N, 180E/1,500 km).

Manatum Tessera (4S, 64E/3,800 km), the second largest of Venus' named tesserae, occupies the far western equatorial part of this region, forming a western extension of the giant Venusian continent Aphrodite Terra. Topographically, Manatum is shaped somewhat like a ring doughnut, with elevated margins which rise to 3,000 m above a relatively depressed central area which lies at mean surface

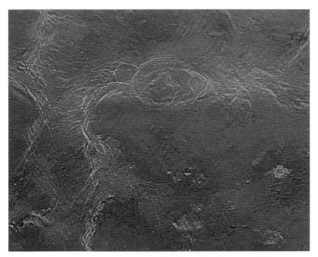

Cauteovan Corona. NASA/
Magellan.

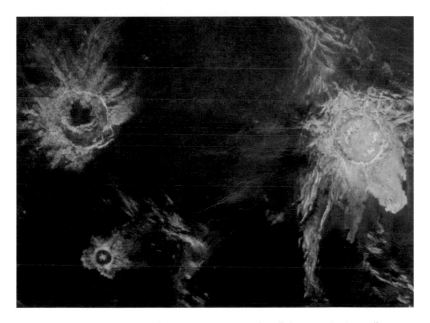

Impact craters Maria Celeste, Greenaway and Callirhoe. NASA/Magellan.

level, possibly formed by mantle downwelling or subsidence due to lithospheric cooling. Manatum is at is broadest and highest in the north, east and south, while a triangular plateau forms its western heights. Central Manatum is presided over by **Verdandi Corona** (6S, 65E/180 km), a well-defined circular feature with a dark smooth interior, crossed east-west by a bundle of fractures.

Manatum's edges are clearly defined in the north by the sharp 2,000 m high scarp **Hestia Rupes** (6N, 71E/588 km). Beyond lies an unnamed plain containing **H'uraru Corona** (9N, 68E/150 km) with its radiating set of fractures, and the young impact crater **Adivar** (9N, 76E/30 km) which, in addition to a set of lobate ejecta flows, is surrounded by a remarkably large radar-light ejecta coma which brushes

The arachnoid Ituana Corona and Praurime Fluctus. NASA/Magellan.

westwards across the radar-dark plains. Several long, prominent dorsa cross the plain yet further north, including **Lemkechen Dorsa** (19N, 69E/2,000 km) and **Unelanuhi Dorsa** (12N, 87E/2,600 km).

The eastern border of Manatum is defined by the rim of **Kaltash Corona** (1N, 75E/450 km), a feature whose northern wall may have been opened up by crustal rifting. Another rift feature, **Tawera Vallis** (12S, 68E/500 km) forms an 80 km wide boundary to the southeast, separating Manatum from the large mountainous area of **Ovda Regio** (3S, 86E/5,280 km), the central upland component of Aphrodite Terra.

With an area of around 10 million square kilometres, Ovda Regio is about the same size as the United States. Most of Ovda is taken up by a tesserated highland plateau averaging 5,000 m high, surrounded by a 'continental shelf' averaging some 2,000 m high and 150 km wide. While many tectonic features in Ovda generally trend east-west, indicating crustal compression from the north and south, its central portions are twisted into a collection of north-south oriented ridges and valleys. Northwestern Ovda is bordered by the sweeping parallel ridges comprising **Nayunuwi Montes** (2N, 83E/900 km), a range formed by tectonic compression,

Close up of the lava plains of Rusalka Planitia. NASA/Magellan.

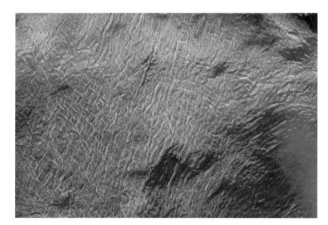

Close-up of the heavily fractured tesserated landscape of Manatum Tessera. NASA/Magellan.

which curve south of **Habonde Corona** (3N, 82E/125 km). To their south lies **Kokomikeis Chasma** (0N, 85E/1,000 km), a smooth-floored lava flooded rift valley averaging 50 km in width.

Southern Ovda contains a large unnamed corona (10S, 89E/100 km), on the southern flanks of which lie the **Lo Shen Valles** (15S, 95E), the planet's most extensive set of lava channels. The valleys meander across an area of some 10,000 square kilometres, most of them winding south into the plains at the edge of **Tahmina Planitia** (23S, 80E/3,000 km).

The tesserated heights of northeastern Ovda bulge northwards towards Sogolon Planitia, where the elongated ridges of Unelanuhi Dorsa generally run parallel to Ovda's coastline at a distance of around 100 km. Several undeformed impact craters in the area, such as **de Beauvoir** (2N, 96E/53 km) indicate the great age of Ovda and the lack of recent major tectonic activity. In the east, overtly tesserated terrain gives way to compressional fractures and dorsa conforming to Ovda's eastern outline.

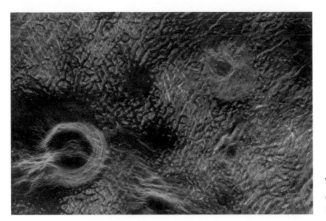

Verdandi Corona in central Manatum Tessera. NASA/Magellan.

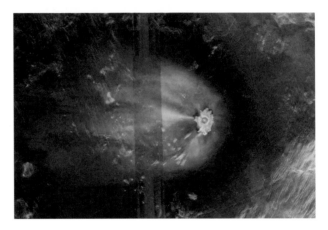

Adivar crater has a remarkable coma-like ejecta. NASA/Magellan.

A number of sizable unnamed lava flooded valleys are to be found in this area, notably one 360 km long and 40 km wide (3N, 106E) and another 450 km long and averaging 65 km wide (2S, 104E), both with smooth, radar dark floors. The north-south trending topography in the area extends towards a marked centre of radiating ridges, fractures and valleys (7S, 108E) at the eastern edge of Ovda Regio. In southern Ovda, two large rift valleys – **Kuanja Chasma** (12S, 100E/890 km) and **Ralk-umgu Chasma** (15S, 106E/840 km) – trend to the east; the latter separates the smooth plains of **Turan Planum** (13S, 117E/800 km) and **Viriplaca Planum** (20S, 112E/1,200 km).

Thetis Regio (11S, 130E/2,801 km) forms the eastern upland component of Aphrodite Terra, most of it lying south of the equator. It has an area of around 5 million square kilometres, around half that of Ovda Regio, comparable with the size of western Europe. Most of Thetis' interior heights are of the order of 4,000 to 5,000 m above mean surface level, but the wider elevated area upon which Thetis rests covers a substantial portion of the planet, from northern **Haasttse-baad Tessera** (6N, 127E/2,600 km) to southern Artemis Corona. Like other Venusian tesserated highland areas, Thetis displays a complicated history of fracturing and folding due to tectonic movements and crustal stresses originating in underlying

Huraru Corona. NASA/Magellan.

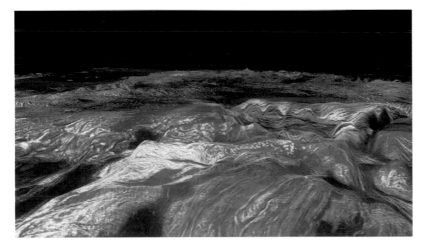

A simulated 3D view of Ovda Regio, looking southward from its northern edge. NASA/Magellan.

mantle movements. In the general pattern of terrain, there is some evidence that the area has undergone rotation in a clockwise fashion, with northeast-southwest trending strike-slip faults in the southeast, west-east faults in the surrounding **Vir-ava Chasma** (17S, 124E/1,700 km) in the south of Thetis, and compressional features to the northeast.

In addition to Turan Planum and Viriplaca Planus, numerous large well-defined rafts of smooth terrain can be found between southern Thetis and the giant Artemis Corona. Artemis, by far the biggest coronal formation on Venus with a diameter of 2,600 km, is a circular feature whose eastern and southern border is clearly

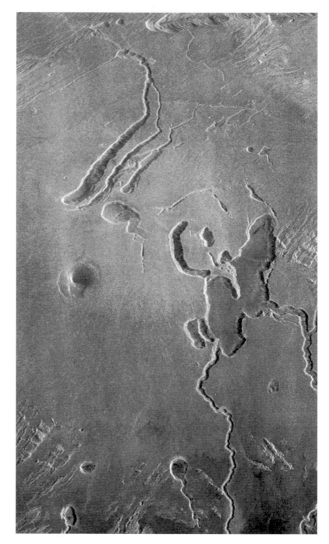

The Lo Shen Valles.
NASA/Magellan.

outlined by Artemis Chasma, a near-continuous peripheral valley more than 3,000 km long, 140 km wide and 2,000 m deep. In the north, Artemis Chasma branches off to link with the zigzagging Quilla Chasma and the northern end of **Britomartis Chasma** (33S, 130E/1,000 km). Britomartis Chasma winds across the centre of Artemis Corona, forking out near Artemis' western edge; one component continues west into Juno Chasma and another ventures south to rejoin Artemis Chasma in the south.

The heights of eastern Thetis narrow and extend north of a smooth unnamed plateau (15S, 140E/950 km) of a similar nature to Ishtar Terra's Lakshmi Planum (see Region One, above). A number of clear-cut graben valleys and rift features are to be found in this area, including **Veden-Ema Vallis** (15S, 141E/300 km), Diana Chasma and **Dali Chasma** (18S, 167E/2,077 km), forming a complex network of valleys. More than 900 km long, Diana Chasma extends east to the nova **Miralaidji**

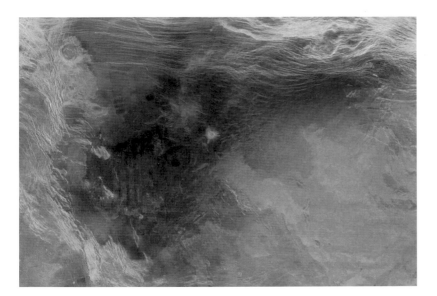

Viriplaca Planum. NASA/Magellan.

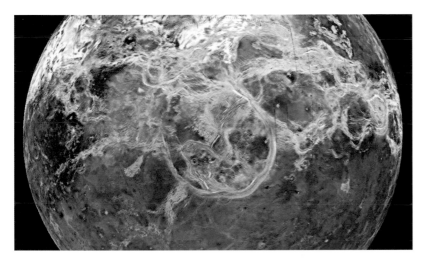

Artemis Corona, Venus' biggest corona. NASA/Magellan.

Corona (14S, 164E/300 km), while Dali Chasma joins the prominent valley-outlined **Atahensik Corona** (19S, 170E/700 km). Yet more deep, prominent valleys stretch northeast from Atahensik Corona towards **Sith Corona** (10S, 177E/350 km) and **Žemina Corona** (12S, 186E/530 km) (see Region Three, below) and south to **Agraulos Corona** (28S, 166E/170 km).

Moving to the southwestern part of this region, Tahmina Planitia extends from Region One (see above), underlying both Manatum Tessera and Ovda Regio. Along Tahmina's southern plains can be found the narrow north-south strip of **Xi Wang-**

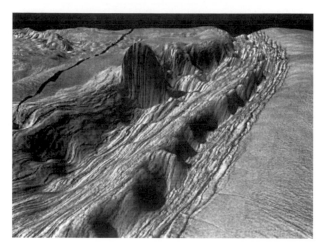

A simulated 3D view of southeastern Artemis Chasma. NASA/Magellan.

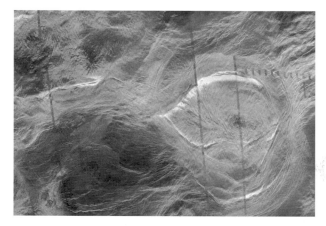

Dali Chasma and Atahensik Corona. NASA/Magellan.

mu Tessera (30S, 62E/1,300 km) and the well-formed coronae **Nishtigri Corona** (25S, 72E/275 km), **Aramaiti Corona** (26S, 82E/350 km) and **Ohogetsu Corona** (27S, 86E/175 km), each of which has a floor sunk deeply below the surrounding plains. A lone topographical highspot of the area is provided by the volcanic cone of **Kunapipi Mons** (34S, 86E/220 km) which rises to 3,000 metres.

Adjoining Tahmina Planitia to its south, **Aino Planitia** (41S, 95E/4,985 km) descends into a fairly smooth basin whose southwestern part contains a cluster of sizeable coronae, notably **Khotun Corona** (47S, 82E/200 km), **Iang-Mdiye Corona** (47S, 86E/300 km), **Cailleach Corona** (48S, 88E/125 km) and **Makh Corona** (49S, 85E/200 km). Aino Planitia's southwestern border is marked by a broad elevated region containing mixed group of features, including the nova **Copia Corona** (43S, 76E/500 km), source of the lava flows of **Ilaheva Fluctus** (43S, 84E/900 km) which blanket parts of Aino, **Zimcerla Dorsa** (48S, 74E/850 km), **Oshumare Dorsa** (59S, 79E/550 km) and **Dunne-Musun Corona** (60S, 85E/630 km).

Xaratanga Chasma, a rift valley system some 1,300 km long, and **Geyaguga Chasma** (57S, 70E/800 km) run along the northern edge of **Mugazo Planitia** (69S, 60E/1,500 km). Impact craters show up well on the radar-dark plains of Mugazo. A remarkable set of roughly even-spaced impact craters, all dotted

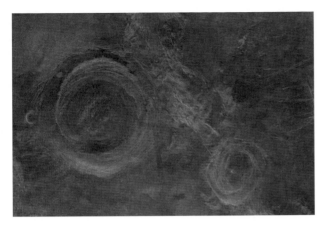

Aramaiti Corona and Ohogetsu Corona in southern Tahmina Planitia. NASA/Magellan.

along the same latitude zone is to be found here, namely **Marsh** (64S, 47E/48 km), **Berggolts** (64S, 53E/30 km), **Danute** (64S, 57E/12 km), **Rand** (64S, 60E/24 km), **Sartika** (64S, 67E/19 km), **Lucia** (62S, 68E/16 km) and **Jitka** (62S, 71E/13 km).

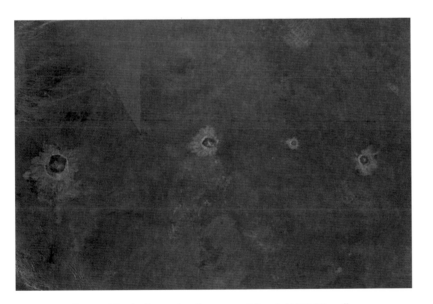

Craters Marsh, Berggolts, Danute and Rand. NASA/Magellan.

The somewhat bland expanse of **Laimdota Planitia** (58S, 117E/1,800 km) merges with its neighbouring planitiae to the northeast and southeast. Of particular note is the impact crater **Addams** (56S, 99E/87 km), whose asymmetric ejecta blanket lends it to the appearance of a mermaid simulacrum. A ridge to the east of Laimdota is occupied by **Sunna Dorsa** (53S, 134E/500 km) and split by **Tellervo Chasma** (60S, 125E/600 km), while **Latmikaik Corona** (64S, 123E/500 km) and **Deohako**

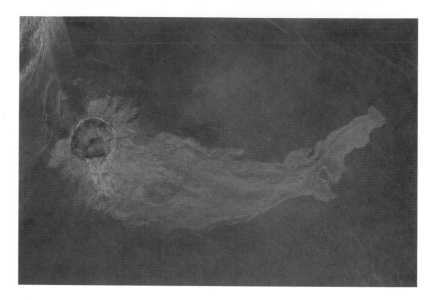

Addams crater and its associated ejecta blanket in Laimdota Planitia. NASA/Magellan.

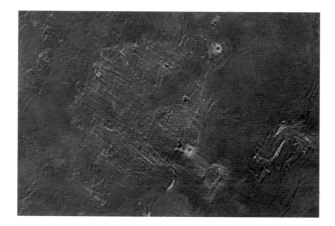

Dsonkwa Regio. NASA/Magellan.

Corona (68S, 118E/300 km) rise to its south. Numerous fluctus lie in the area, notable among them **Arubani Fluctus** (55S, 132E/620 km) and **Nambubi Fluctus** (61S, 135E/850 km).

Few prominent topographical features are to be found in the remaining far southeastern section of this region, which includes **Zhibek Planitia** (40S, 157E/2,000 km), **Imapinua Planitia** (60S, 142E/2,100 km) and the western reaches of **Nsomeka Planitia** (53S, 195E/2,100). An uplift at the junction of these three plains forms **Dsonkwa Regio** (53S, 167E/1,500 km), a hilly area presided over by **Tonatzin Corona** (53S, 164E/400 km), crossed by **Nortia Tesserae** (49S, 160E/650 km) in the north and by the striated landscape of **Mena Colles** (53S, 160E/850 km) in the west. Finally, the plains of **Alma-Merghen Planitia** (76S, 100E/1,500 km) extend southward towards the pole.

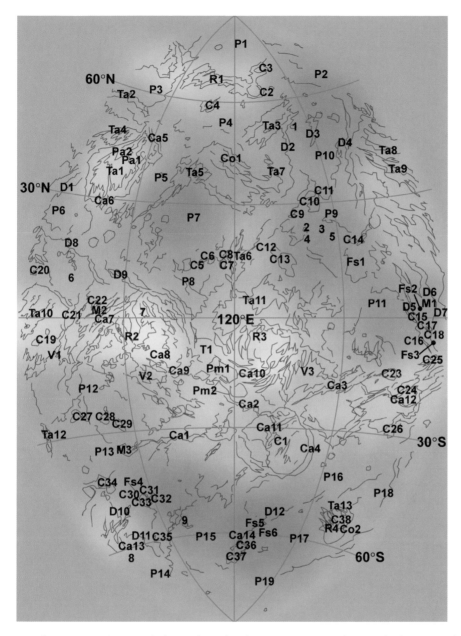

Map of Region Two (60-180E), the Niobe-Aphrodite-Artemis Region, showing the locations of features described in the text.

Key to Features on Chart:

Pa Patera; Rs Rupes; Ta Tessera; Ca Chasma; Fo Fossa; Fs Fluctus; Fa Farra; Co Colles; M Mons; Pm Planum; P Planitia; C Corona; D Dorsa; T Terra; V Vallis; R Regio; C1. Artemis Corona; C2. Nightingale Corona; C3. Earhart Corona; C4. Fakahotu Corona; C5. Dhisana Corona; C6. Allatu Corona; C7. Bhumiya Corona; C8. Omeciuatl Corona; C9. Boann Corona; C10. Cauteovan Corona; C11. Ved-Ava Corona; C12. Abundia Corona; C13. Kubebe Corona; C14. Ituana Corona; C15. Hannahannas Corona; C16. Nirmali Corona; C17. Saunau Corona; C18. Eigin Corona; C19. Verdandi Corona; C20. H'uraru Corona; C21. Kaltash Corona; C22. Habonde Corona; C23. Miralaidji Corona; C24. Atahensik Corona; C25. Sith Corona; C26. Agraulos Corona; C27. Nishtigri Corona; C28. Aramaiti Corona; C29. Ohogetsu Corona; C30. Khotun Corona; C31. Iang-Mdiye Corona; C32. Cailleach Corona;

Region Three: The Kawelu-Atla-Helen Region (180 to 300E)

This third region of Venus is best described as one comprising an assortment of dorsa-wrinkled planitiae and chasma-rifted, mountain-punctuated regiones. Of the patchwork of planitiae in northwest, **Vinmara Planitia** (54N, 208E/1,635 km) and **Ganiki Planitia** (40N, 202E/5,160 km) are the most heavily wrinkled, with long ridges that link small areas of tesserated terrain, while eastern **Kawelu Planitia** (33N, 247E/3,910 km), **Libuše Planitia** (60N, 290E/1,200 km) and the far western reaches of Guinevere Planitia (also see Region One, above) are considerably smoother. To the south, a series of regiones – from **Atla Regio** (9N, 200E/3,200 km) in the west across to **Beta Regio** (25N, 283E/2,869 km) in the east – are linked together by a huge network of chasma, from Dali Chasma (see also Region Two, above), through **Zewana Chasma** (9N, 212E/900 km) and **Hecate Chasma** (18N, 254E/3,145 km) to **Devana Chasma** (16N, 285E/4,600 km).

The smooth far northern lowland plain of **Snegurochka Planitia** (87N, 328E/2,775 km) is bordered by **Dennitsa Dorsa** (86N, 206E/872 km) to the west. Dennitsa Dorsa is a far northern extension of a north-south trending system of dorsa which cross Vinmara Planitia, one of the most wrinkled areas on the entire planet. The most prominent among these features include **Lukelong Dorsa** (73N, 179E/1,566 km), **Lauma Dorsa** (65N, 190E/1,517 km), **Ahsonnutli Dorsa** (48N, 197E/1,708 km), **Pandrosos Dorsa** (58N, 208E/1,254 km). Pandrosos Dorsa stretches between two clusters of coronae – **Muzamuza Corona** (66N, 205E/163 km), **Cassatt Corona** (66N, 208E/152 km) and **Nzingha Corona** (69N, 206E/140 km) in the north and **Cerridwen Corona** (50N, 202E/217 km) and **Neyterkob Corona** (50N, 205E/211 km) in the south.

C33. Makh Corona; C34. Copia Corona; C35. Dunne-Musun Corona; C36. Latmikaik Corona; C37. Deohako Corona; C38. Tonatzin Corona; Pm1. Turan Planum; Pm2. Viriplaca Planum; P1. Louhi Planitia; P2. Atalanta Planitia; P3. Audra Planitia; P4. Tilli-Hanum Planitia; P5. Lowana Planitia; P6. Akhtamar Planitia; P7. Niobe Planitia; P8. Sogolon Planitia; P9. Llorona Planitia; P10. Vellamo Planitia; P11. Rusalka Planitia; P12. Tahmina Planitia; P13. Aino Planitia; P14. Mugazo Planitia; P15. Laimdota Planitia; P16. Zhibek Planitia; P17. Imapinua Planitia; P18. Nsomeka Planitia; P19. Alma-Merghen Planitia; M1. Lamashtu Mons; M2. Nayunuwi Montes; M3. Kunapipi Mons; R1. Tethus Regio; R2. Ovda Regio; R3. Thetis Regio; R4. Dsonkwa Regio; Co1. Akkruva Colles; Co2. Mena Colles; T1. Aphrodite Terra; Ta1. Tellus Tessera; Ta2. Dekla Tessera; Ta3. Ananke Tessera; Ta4. Meni Tessera; Ta5. Kutue Tessera; Ta6. Gegute Tessera; Ta7. Likho Tesserae; Ta8. Nemesis Tesserae; Ta9. Athena Tessera; Ta10. Manatum Tessera; Ta11. Haasttse-baad Tessera; Ta12. Xi Wang-mu Tessera; Ta13. Nortia Tesserae; Ca1. Juno Chasma; Ca2. Quilla Chasma; Ca3. Diana Chasma; Ca4. Artemis Chasma; Ca5. Medeina Chasma; Ca6. Kottravey Chasma; Ca7. Kokomikeis Chasma; Ca8. Kuanja Chasma; Ca9. Ralk-umgu Chasma; Ca10. Vir-ava Chasma; Ca11. Britomartis Chasma; Ca12. Dali Chasma; Ca13. Geyaguga Chasma; Ca14. Tellervo Chasma; Rs1. Hestia Rupes; V1. Tawera Vallis; V2. Lo Shen Valles; V3. Veden-Ema Vallis; Pa1. Apgar Patera; Pa2. Eliot Patera; D1. Mardezh-Ava Dorsa; D2. Nephele Dorsa; D3. Frigg Dorsa; D4. Vedma Dorsa; D5. Zaryanitsa Dorsa; D6. Yalyane Dorsa; D7. Poludnitsa Dorsa; D8. Lemkechen Dorsa; D9. Unelanuhi Dorsa; D10. Zimcerla Dorsa; D11. Oshumare Dorsa; D12. Sunna Dorsa; Fs1. Praurime Fluctus; Fs2. Argimpasa Fluctus; Fs3. Dotetem Fluctus; Fs4. Ilaheva Fluctus; Fs5. Arubani Fluctus; Fs6. Nambubi Fluctus; 1. Cochran; 2. Maria Celeste; 3. Greenaway; 4. Callirhoe; 5. Bourke-White; 6. Adivar; 7. de Beauvoir; 8. Jitka; 9. Addams

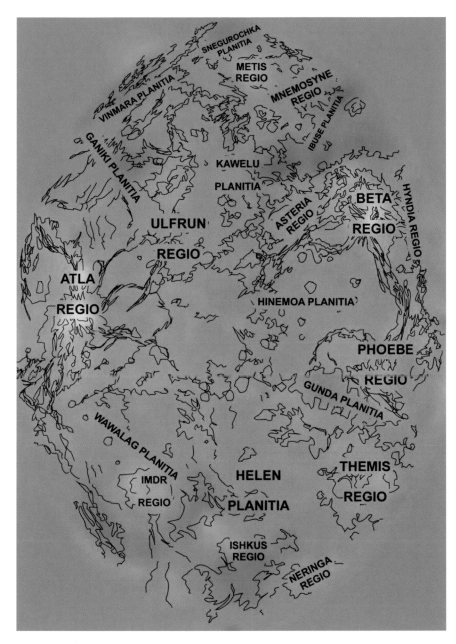

Chart showing the main areas in Region Three, from 180 to 300 east longitude.

Metis Regio (71N, 252E/729 km) rises to the south of Snegurochka Planitia, from which **Thallo Mons** (76N, 234E/216 km) and **Renpet Mons** (76N, 236E/138 km) stand out. The lower but hillier parts of eastern Metis blends into **Mnemosyne Regio** (66N, 280E/875 km), an area containing the nova **Feronia Corona** (68N, 282E/360 km), **Coatlicue Corona** (63N, 273E/199 km) and **Rananeida Corona** (63N, 264E/448 km). In eastern Mnemosyne, the impact crater **Duncan** (68N, 292E/40 km) stands out prominently on a radar-dark plain fractured by numerous linear faults.

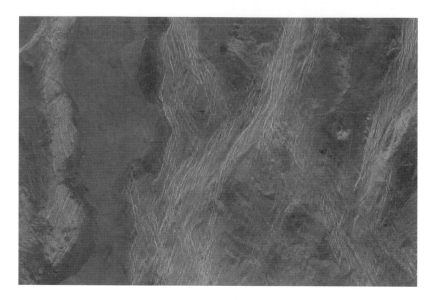

Pandrosos Dorsa in Vinmara Planitia. NASA/Magellan.

Okipeta Dorsa (68N, 240E/1,200 km) on Metis' southern border runs along the edge of an unnamed straight flooded rift valley (66N, 245E) some 1,090 km long and averaging 120 km in breadth. To the south, the irregular patches of **Virilis Tesserae** (56N, 240E/782 km) are sharply defined in the southwest, making a northwest-southeast line adjacent to the smooth northwestern reaches of Kawelu Planitia.

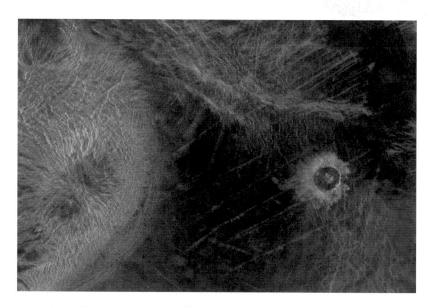

Nova Feronia Corona (at left) and the crater Duncan. NASA/Magellan.

Okipeta Dorsa run along the northern edge of a large unnamed flooded rift valley. NASA/Magellan.

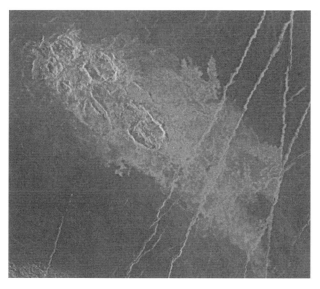

Terhi (46N, 253E/11 km), an irregular crater in southern Kawelu Planitia.

Ganiki Planitia, a heavily wrinkled area, stretches from **Vellamo Planitia** (45N, 149E/2,155 km) across to Ulfrun Regio. The tesserated patches of **Nemesis Tesserae** (40N, 181E/355 km) and **Lahevhev Tesserae** (29N, 189E/1,300 km) spring up in Ganiki, along with the linear cliffs of **Fornax Rupes** (30N, 201E/729 km).

Ulfrun Regio at the eastern border of Ganiki rises in a large north-south trending series of ridges separated by numerous fossae, from **Bellona Fossae** (38N, 222E/855 km) in the north, near **Sakwap-mana Mons** (35N, 220E/500 km), to **Fea Fossae** (28N, 224E/620 km) and **Zisa Corona** (12N, 221E/850 km). Ulfrun forms a sort of junction between two huge chasmata systems – Zewana Chasma in Atla Regio to its west and Hecate Chasma in **Asteria Regio** (22N, 268E/1,131 km) to its east.

From the smooth plains of Kawelu Planitia, northeast of Ulfrun Regio, arise several large volcanic mounds, notably **Sekmet Mons** (45N, 241E/285 km), **Venilia Mons** (33N, 239E/320 km) and **Atira Mons** (52N, 268E/152 km) on the border with western Guinevere Planitia (see Region One, above). Kawelu's transition into Asteria Regio to its south is marked by the hilly west-east oriented tesserated terrain of **Sudenitsa Tesserae** (33N, 270E/4,200 km) and the north-south oriented **Yuki-Onne Tessera** (39N, 261E/1,200 km). Southern Asteria is rent by Hecate Chasma, a 3,145 km long rift valley and ridge system caused by mantle upwelling and crustal tension, with indications of tectonic boundary spreading. Hecate extends between the nova **Taranga Corona** (17N, 252E/525 km) and **Polik-mana Mons** (25N, 264E/600 km).

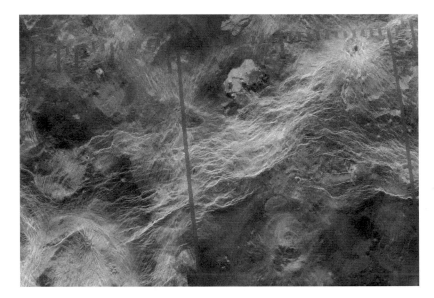

Hecate Chasma. NASA/Magellan.

Adjoining Asteria Regio and continuing its systems of chasmata and tesserated terrain, lies Beta Regio, dominated by its radar-bright central feature **Theia Mons** (23N, 281E/226 km). Theia Mons, one of Venus' largest volcanoes, rises to a height of more than 4,000 m and is split longitudinally by the huge Devana Chasma which extends north-south for some 4,600 km, cutting northwards into **Rhea Mons** (32N, 282E/217 km) and southwards into the neighbouring uplands of **Phoebe Regio** (6S, 283E/2,852 km). Western Beta Regio also contains **Žverine Chasma** (19N, 271E/1,300 km) and **Latona Chasma** (26N 268E/530 km), the latter being an extension of Hecate Chasma east of Polik-mana Mons. Eastern Beta Regio is adjoined by **Hyndla Regio** (23N, 295E/2,300 km), a narrow north-south upland ridge which bridges the tesserated terrain of **Zirka Tessera** (33N, 300E/450 km) and **Nedolya Tesserae** (5N, 294E/1,200 km).

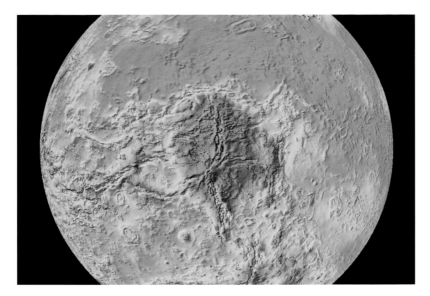

The huge volcanic highland of Beta Regio, crowned by the Theia Mons, dominates its surroundings. NASA/Magellan.

Atla Regio spans the equator in the western part of this region. Atla has many similarities to Beta Regio – it is a major uplifted area containing a number of large volcanoes, cut through by several deep and extensive chasmata. **Maat Mons** (1N, 195E/395 km), a giant volcano straddling the equator, rises to 1,700 m above the surrounding mountainous terrain, its summit being some 8,000 m above mean Venusian surface level. Immediately to its south are the jumbled heights of **Ongwuti Mons** (2S, 195E/500 km) which lie in a heavily rifted unnamed zone along a southern elevated extension of Atla Regio. North of Maat Mons, **Ganis Chasma** (15N, 194E/615 km) cleaves an elevated region from **Ozza Mons** (5N, 201E/507 km) to **Nokomis Montes** (20N, 189E/486 km). To the west, **Sapas Mons** (9N, 188E/217 km), an isolated volcano with a caldera 25 km across, rises to 1,500 m above its hilly surroundings in eastern Rusalka Planitia (also see Region Two, above). Sapas lies at the centre of a striking starburst of radar bright fractures and lava flows set against a radar-dark landscape. The eastern plain of Rusalka also features **von Schuurman** (5S, 191E/29 km), an impact crater immediately surrounded by radar-bright flow lobes amid a much larger corona of light ejecta which has been brushed westwards.

Several chasmata converge in the vicinity of Ozza Mons, namely **Tkashi-mapa Chasma** (13N, 206E/1,100 km) from the north, which cuts through the eastern flanks of the corona **Nahas-tsan Mons** (14N, 205E/500 km), Zewana Chasma from the northeast and **Kicheda Chasma** (3S, 213E/1,500 km) from the east. South of the latter can be found a variety of features, including **Ningyo Fluctus** (6S, 206E/970 km), nova **Oduduwa Corona** (11S, 212E/150 km) and the coronae **Jotuni Patera** (7S, 214E/80 × 104 km) and **Maram Corona** (8S, 222E/600 km).

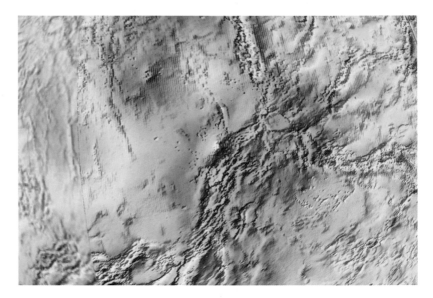

Topographic chart of Atla Regio centred on Maat Mons. NASA/Magellan.

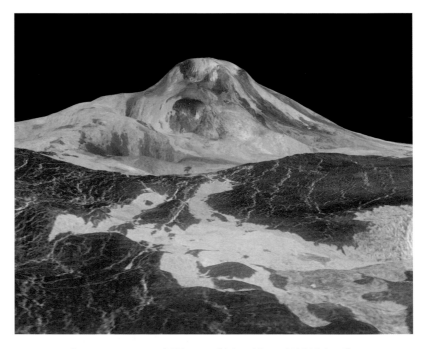

Computer-generated 3D view of Maat Mons. NASA/Magellan.

Computer-generated 3D view of Sapas Mons. NASA/Magellan.

Sapas Mons. NASA/Magellan.

The lowlands of **Hinemoa Planitia** (5N, 265E/3,700 km) extend across the central equatorial part of this region. **Aruru Corona** (9N, 262E/450 km), a volcanic uplift with a distorted outline, stands out along with the adjacent **Lama Tholus** (8N, 266E/110 km) in northeastern Hinemoa. To the east lie the isolated volcanoes **Tuulikki Mons** (10N, 275E/520 km) and **Xochiquetzal Mons** (4N, 270E/80 km). **Chimon-mana Tessera** (3S, 270E/1,500 km) forms a curving swathe from Phoebe Regio in the far east to **Uretsete Mons** (12S, 261E/500 km) in southeastern Hinemoa.

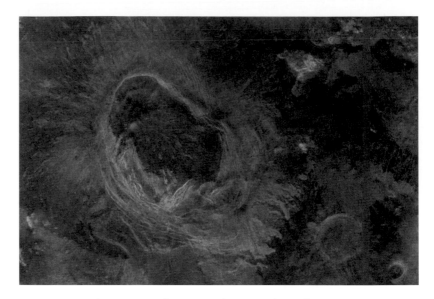

Aruru Corona, one of Hinemoa Planitia's volcanic features. NASA.

Much of central and southwestern Hinemoa is slightly elevated and occupied by sizeable coronae, including the heart shaped nova **Javine Corona** (6S, 251E/450 km), nova **Dhorani Corona** (8S, 243E/200 km), **Erkir Corona** (16S, 234E/275 km), **Atete Corona** (16S, 244E/600 km) and **Ludjatako Corona** (13S, 251E/300 km).

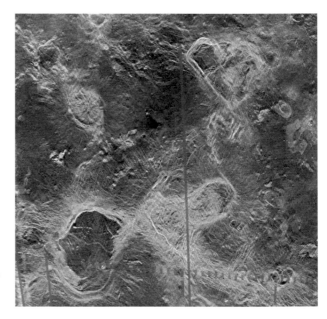

Coronal cluster in southern Hinemoa Planitia, including Atete Corona at lower left. NASA/Magellan.

Large tracts of western **Wawalag Planitia** (30S, 217E/2,600 km), a large plain stretching between southern Atla Regio and **Imdr Regio** (43S, 212E/1,611 km), are wrinkled by the north-south trending ridges of **Aditi Dorsa** (30S, 189E/1,200 km) and **Sirona Dorsa** (44S, 194E/700 km). **Isabella** (30S, 204E/175 km), the planet's second largest impact crater, stands out prominently in central Wawalag. Two large flow lobes extending to the south and southeast of Isabella, the southern flow partially engulfing the flanks of the arachnoid **Nott Corona** (32S, 202E/150 km) while the southeastern flow is overlain by ejecta from the younger impact crater **Cohn** (33S, 208E/18 km).

Imdr Regio, an oval-shaped highland plateau with an area of around 850,000 square kilometres, is a volcanic uplift on the southern edge of Wawalag Planitia. It is crossed by a number of ridges, including **Nuvakchin Dorsa** (53S, 212E/2,200 km) and **Arev Dorsa** (52S, 216E/420 km) in the south, which skirt the flanks of **Idunn Mons** (47S, 215E/250 km) Imdr's highest feature with a summit that rises more than 3,000 m above the surrounding plateau. A large valley system, **Olapa Chasma** (42S, 209E/650 km) extends northeast from the flanks of Idunn Mons.

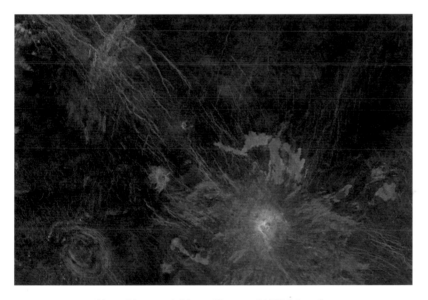

Idunn Mons and Olapa Chasma. NASA/Magellan.

Half a dozen wrinkle ridges extend across eastern Nsomeka Planitia (see also Region Two, above) into **Nuptadi Planitia** (73S, 250E/1,200 km), the largest of which include Nuvakchin Dorsa and **Rokapi Dorsa** (55S, 222E/2,200 km). Nuptadi itself extends south towards Venus' south pole. To its north lie two small wrinkled regiones, **Ishkus Regio** (61S, 245E/1,000 km) and **Neringa Regio** (65S, 288E/1,100 km); Ishkus is more elevated than Neringa, and boasts its own sizeable volcano, **Awenhai Mons** (60S, 248E/100 km).

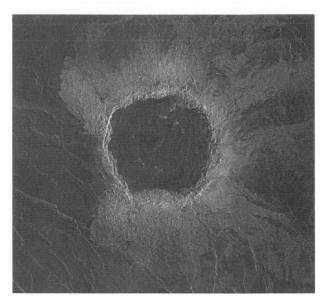

Sayers (68S, 230E/98 km) a polygonal flooded crater in northern Nuptadi Planitia. NASA/Magellan.

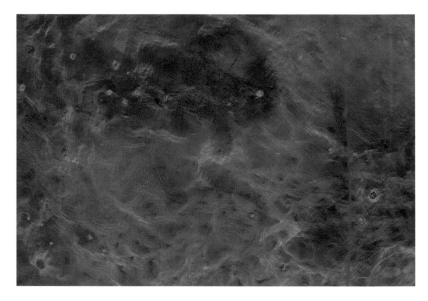

The wrinkled plains of Ishkus Regio (at left) and Neringa Regio. NASA/Magellan.

Helen Planitia (52S, 264E/4,360 km), Venus' third largest planitia, takes up much of the central southern part of this region. Several coronae lie dotted around the area, from Oanuava Coronae (33S, 256E/375 km) in the north to Naotsete Corona (58S, 250E/200 km) in the south. The lowland basin in western Helen is traversed by a number of north-south ridges, including Tinianavyt Dorsa (51S, 239E/1,500 km) and Kastiatsi Dorsa (53S, 245E/1,200 km), compressional features that appeared in a region of mantle downwelling.

Finally, in the southeast lies Themis Regio (37S, 284E/1,811 km), an extensive upland region rifted by a series of chasmata and bulging with numerous large

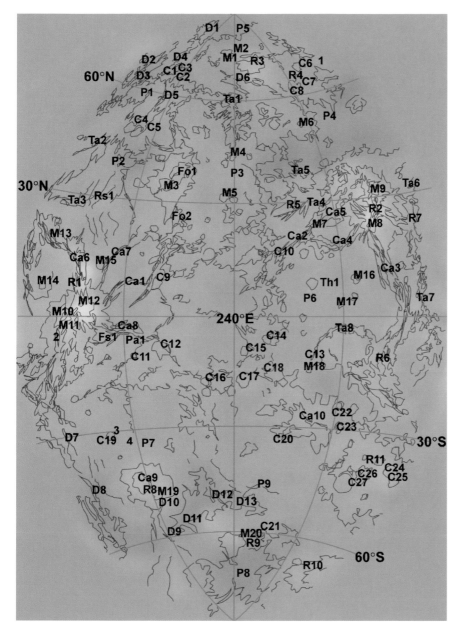

Map of Region Three (180-300E), the Kawelu-Atla-Helen Region, showing the locations of features described in the text.

Key to features on chart:

Pa Patera; Rs RupesA©; Ta Tessera; Ca Chasma; Fo Fossa; Fs Fluctus; Fa Farra; Co Colles; M Mons; Pm Planum; P Planitia; C Corona; D Dorsa; T Terra; V Vallis; R Regio; C1. Muzamuza Corona; C2. Cassatt Corona; C3. Nzingha Corona; C4. Cerridwen Corona; C5. Neyterkob Corona; C6. Feronia Corona; C7. Coatlicue Corona; C8. Rananeida Corona; C9. Zisa Corona; C10. Taranga Corona; C11. Oduduwa Corona; C12. Maram Corona; C13. Aruru Corona; C14. Javine Corona; C15. Dhorani Corona; C16. Erkir Corona; C17. Atete Corona; C18. Ludjatako Corona; C19. Nott Corona; C20. Oanuava Coronae; C21. Naotsete Corona; C22. Hervor Corona; C23. Lilwani Corona; C24. Semiramus Corona; C25. Ukemochi Corona; C26. Shulamite Corona; C27. Shiwanokia Corona; Th1. Lama Tholus; P1. Vinmara Planitia; P2. Ganiki Planitia; P3. Kawelu Planitia; P4. Libuše Planitia; P5. Snegurochka Planitia; P6.

coronal uplifts. **Parga Chasmata** (20S, 255E/11,000 km, Venus' longest chasma) cuts through in northwestern Themis, extending south of **Hervor Corona** (26S, 269E/250 km) and **Lilwani Corona** (30S, 272E/500 km), across to **Semiramus Corona** (37S, 293E/375 km) and **Ukemochi Corona** (39S, 296E/300 km). Southern Themis boasts two of Venus' loftiest coronal uplifts, namely the neighbouring novae **Shulamite Corona** (39S, 284E/275 km) and **Shiwanokia Corona** (42S, 280E/500 km). Several isolated volcanoes peek out from the rough terrain in far eastern Themis, notably **Tefnut Mons** (39S, 304E/182 km) and **Faravari Mons** (44S, 309E/500 km).

Hinemoa Planitia; P7. Wawalag Planitia; P8. Nuptadi Planitia; P9. Helen Planitia; D1. Dennitsa Dorsa; D2. Lukelong Dorsa; D3. Laūma Dorsa; D4. Ahsonnutli Dorsa; D5. Pandrosos Dorsa; D6. Okipeta Dorsa; D7. Aditi Dorsa; D8. Sirona Dorsa; D9. Nuvakchin Dorsa; D10. Arev Dorsa; D11. Rokapi Dorsa; D12. Tinianavyt Dorsa; D13. Kastiatsi Dorsa; Ta1. Virilis Tesserae; Ta2. Nemesis Tesserae; Ta3. Lahevhev Tesserae; Ta4. Sudenitsa Tesserae; Ta5. Yuki-Onne Tessera; Ta6. Zirka Tessera; Ta7. Nedolya Tesserae; Ta8. Chimon-mana Tessera; Fs1. Ningyo Fluctus; Pa1. Jotuni Patera; Rs1. Fornax Rupes; Fo1. Bellona Fossae; Fo2. Fea Fossae; R1. Atla Regio; R2. Beta Regio; R3. Metis Regio; R4. Mnemosyne Regio; R5. Asteria Regio; R6. Phoebe Regio; R7. Hyndla Regio; R8. Imdr Regio; R9. Ishkus Regio; R10. Neringa Regio; R11. Themis Regio; Ca1. Zewana Chasma; Ca2. Hecate Chasma; Ca3. Devana Chasma; Ca4. žverine Chasma;p Ca5. Latona Chasma; Ca6. Ganis Chasma; Ca7. Tkashi-mapa Chasma; Ca8. Kicheda Chasma; Ca9. Olapa Chasma; Ca10. Parga Chasmata; M1. Thallo Mons; M2. Renpet Mons; M3. Sakwap-mana Mons; M4. Sekmet Mons; M5. Venilia Mons; M6. Atira Mons; M7. Polik-mana Mons; M8. Theia Mons; M9. Rhea Mons; M10. Maat Mons; M11. Ongwuti Mons; M12. Ozza Mons; M13. Nokomis Montes; M14. Sapas Mons; M15. Nahas-tsan Mons; M16. Tuulikki Mons; M17. Xochiquetzal Mons; M18. Uretsete Mons; M19. Idunn Mons; M20. Awenhai Mons; 1. Duncan; 2. von Schuurman; 3. Isabella; 4. Cohn

Observing Mercury and Venus

Chapter 4

Equipment for Observing Mercury and Venus

Vision

A small but powerful pair of binoculars called 'eyes' is the most important optical equipment belonging to any visual observer, and it's important to treat these precious little instruments with care in order to maximise your enjoyment of visual astronomical observation.

Eye Care

It's always advisable for mature folk to have annual eye checkups, as some unsuspected but treatable medical conditions (including non-visual problems) may come to light as a result.

Smoking tobacco is detrimental to astronomical observation, let alone the smoker's general health. Our eyes use nearly all of the oxygen that reaches them through the blood vessels; rather worryingly, smoking reduces blood oxygen content by around 10 percent per pack smoked per day, since the carbon monoxide in cigarette smoke attaches to the haemoglobin in red blood cells more readily than oxygen itself.

Drinking alcohol while observing decreases how much detail can be seen on Mercury and Venus in direct proportion to the amount of alcohol consumed, and the observer's efficiency is totally compromised when, despite the observer's best efforts, the eye cannot be made to remain in proximity to the eyepiece. Alcohol dilates the blood vessels, and though it may make the consumer feel warm for a short while, the extra heat loss from the body can prove dangerous on cold nights. So, alcohol ought to be avoided until after the observing session when one is safely indoors at home.

Foods beneficial to healthy vision are a variety of fruits and vegetables, especially dark-coloured ones like carrots and broccoli that are good sources of beta-carotene and many carotenoids. These substances aid night vision and help maintain good vision. Vitamin C helps protect the eyes against ultraviolet radiation, and being an antioxidant it inhibits the natural oxidisation of cells. The development of cataracts and age-related macular degeneration (AMD) is inhibited by Vitamin E, a substance found in wheat germ oil, sunflower seed and oil, hazelnuts, almonds, wheat germ, fortified cereals and peanut butter. Finally, zinc helps maintain a healthy retina and can play a part in preventing AMD. Zinc is found in wheat germ, sunflower seeds, almonds, tofu, brown rice, milk, beef and chicken. Visual acuity actually diminishes if blood sugar levels are low, so a small snack – be it a chocolate bar or a banana – during the observing session is both pleasant and beneficial. But don't throw the banana peel on the ground, otherwise you might end up seeing more stars than you'd bargained for.

Solar Folly – a Warning

When making certain observations of Mercury and Venus – particularly observations of transits across the Sun or sweeping for them using binoculars or telescopes during the daytime – the Sun's intense light poses a potential danger to the eyes. Great care must be taken never to allow direct sunlight to enter the eye through any unfiltered optical instruments, since a split second of unprotected exposure to the Sun will permanently damage the retina and perhaps cause a degree of blindness.

Being a sufferer of permanent retinal damage due to sunlight, I know what I'm talking about. As a keen young stargazer I once fabricated what I foolishly considered to be a solar filter using several dark plastic sunglass lenses sandwiched together and taped to the eyepiece. Using my right eye, the Sun was found easily enough and centred in the field of view; yet within seconds the intense focused sunlight had burned a hole through the hopelessly inadequate filter and had flashed into my eye, temporarily overwhelming my ability to see. Alas, permanent damage had been done. Ever since that foolish experiment in the summer of 1976 I have suffered a permanent patch of visual distortion at the very centre of my right eye's field of view. My right eye is useless for visual observation on its own – Venus looks like an exploding artillery shell and Mercury looks like a misshaped pyrotechnic flash. Not a pretty sight, nor one that can ever be improved.

Ocular Anatomy

Our eyes are spherical organs measuring around 4 cm across. A transparent membrane at the front of the eye called the cornea focuses light rays through a chamber filled with transparent fluid, the aqueous humour, through an opening in the iris called the pupil, and on through a clear crystalline lens behind it. This lens focuses light across a chamber filled with vitreous humour, a clear jelly that gives the eyeball its rigidity, and an upside-down image is projected onto the retina inside the back part of the eyeball. The retina contains millions of light sensitive cells called rods and cones that convert the light into electrical impulses which are sent directly to the image processing centre of the brain through more than one million nerve cells in the optic nerve. The brain automatically flips the image the right way up and processes the image.

A spot called the fovea, which has the greatest concentration of cones, lies at the very centre of the retina, allowing high detail colour vision at the centre of the field of view. There are three types of cone cells – ones sensitive to red, yellow-green and blue light – but the cones are not triggered in low light level situations. Objects like Mercury and Venus are always bright enough to trigger the cones, permitting detailed colour vision, but in low light circumstances – for example when attempting to view dim deep sky objects – only the rod cells lying around the fovea, away from the centre of the field of view, are stimulated.

UV Sensitivity and Venus' Cloud Features

Many observers with excellent eyesight find it difficult to see any detail at all on the Venusian disk, when others with less acute vision claim to be able to clearly discern cloud features. This apparent paradox – which for centuries tended to cast doubt upon the observations of those who could easily discern features on Venus – is explained by the fact that the clouds of Venus are far better seen in ultraviolet (UV) light than in integrated light. Visual sensitivity to UV varies between observers because the eye's lens absorbs these wavelengths, filtering them out before they impact on the retina. While all three cone types in the retina are sensitive to UV light, blue cones are the most UV sensitive. To avoid overloading the blue cones when setting up the telescope prior to observing Venus, a pair of red filtered sunshades may prove a useful accessory. Ironically, cataract surgery can actually improve a person's ability to perceive Venusian cloud detail, since the operation removes the UV filter of the eye's lens.

Averted Vision

To get the best view of a very dim object the observer can take advantage of a technique called averted vision, by looking to one side of the object, stimulating the areas within the retina containing the highest concentration of rods. Averted vision works best when the object being viewed is positioned at an apparent angle of around 15 to 20° away from the centre of the field of view towards one's nose. Therefore, a telescopic observer using their right eye to view an object needs to look to the right side of the target, while a left-eyed viewer will need to look towards the left of the target object to maximise the effects of averted vision. Rod cells do not deliver as detailed images as the cones, nor are they capable of distinguishing between different colours, so most dim deep sky objects appear in shades of grey when viewed visually.

Blind Spot

A lack of photoreceptive cells where the optic nerve enters the retina produces a blind spot in each eye – they lie on the left side of the left eye's field of view, and on the right side of the right eye's field of view. Although the fact that we have a blind spot in each eye doesn't really affect observers of Mercury and Venus, the blind spot's presence may be demonstrated using averted vision. The following experiment demonstrates the complete blindness of the blind spot, and it usually astounds anyone first trying it. Cover your left eye with your left hand, and using your right eye slowly scan the area to the left of Venus, a few degrees away from it. You will eventually locate a spot to look at where Venus has completely vanished from sight. The actual area covered by the blind spot measures about 6° across, allowing Venus to be kept out of sight despite those slight involuntary movements of the eye that happen when viewing a small area for any length of time (see saccades, below).

Floaters

Visible when viewing bright objects, floaters can take the form of minute dark flecks, translucent cobwebs or clouds of various shapes and sizes. They are the shadows cast onto the retina by the remnants of dead cells floating in the vitreous humour. Everyone has floaters, and they can be annoying because they can obscure planetary features. Floaters become increasingly numerous and with age, but most people put up with them. However, if the condition becomes particularly annoying or disruptive to normal vision, laser eye surgery or a vitrectomy can remove them.

Saccades

Every visual planetary observer must contend with involuntary eye movements known as saccades. They consist of swift (velocity of 1,000° per second) near-instantaneous minute jumps of the eyeball from one point to another. Under normal circumstances they occur without the conscious awareness of the observer, the eye involuntarily moving up to three times each second due to saccades. However, saccade activity becomes known to most planetary observers wishing to scrutinise objects with relatively small apparent angular diameters in the telescope eyepiece, such as the illuminated portions of Mercury or Venus. Saccades make it impossible to concentrate exclusively on one particular point on a planetary surface for any great length of time. Despite any amount of conscious undivided attention, the eye will invariably jump to a point close to that being scrutinised within just a few seconds. The effect is exacerbated by the fact that planetary surfaces generally contain subtle, non-overt detail and they do not possess any degree of movement or change that will hold the observer's attention in real time.

It may seem annoying, but saccades serve a number of important purposes. They allow an object under scrutiny to be centred in the most visually sensitive part of the retina, at the fovea, helping the observer build a mental map of an object; saccades also bring an area of interest rapidly back to the centre of the field of view, which in terms of survival is important when following a moving object with minimal conscious head or eye movements. Incredibly, experiments have shown that if saccades were not present, and an object like Mercury or Venus were able to be centred in one's field of view for any length of time, the planetary image would soon disappear from view until the eye was moved slightly.

Monoculars and Binoculars

When it comes to viewing Mercury and Venus, regular monoculars and binoculars are thoroughly capable instruments for certain types of low-magnification observation – close appulses, conjunctions and occultations by the Moon and close encounters with other planets and stars, for example. When steadily held, a regular monocular, pair of opera glasses or binoculars will just about reveal the crescent phases of Venus.

Monoculars and binoculars have numerous advantages over telescopes. They are usually far less expensive than telescopes, they are far less cumbersome to carry around (even with a stand), they deliver a wide field of view and they are more robust than telescopes, capable of retaining their optical collimation after taking occasional knocks.

Monoculars are little hand-held telescopes with small (typically 20–30 mm) achromatic objective lenses; they use roof prisms to deliver low magnification, right way up views. Monoculars can be carried in a coat pocket without any difficulty and they are great for locating Mercury and Venus in circumstances that prove challenging when using the unaided eye alone or for observing appulses between the inferior planets and other celestial objects.

Opera glasses (also known as Galilean binoculars) are the simplest binocular design. In their most basic form, they consist of small biconvex objective lenses and biconcave eye lenses, and these form an upright image. Opera glasses have a short focal length, and produce a low magnification and a narrow field of view. Their unsophisticated optical configuration produces false colour (called chromatic aberration) which becomes obvious when bright objects like Venus are viewed. Often made to fold down to a compact size when not in use, opera glasses are lightweight and easy to carry around in a coat pocket or handbag. Like monoculars, opera glasses are useful to keep in the pocket, since they can be used for quick peeks at the skies, sweeping the horizon for that first view of Mercury or Venus in the dawn or dusk skies.

Binoculars

Binocular power is identified by two figures that denote their magnification and the size of their objective lenses – 7× 30 binoculars give a magnification of 7× and have 30 mm objective lenses. Small to medium sized binoculars (with objective lenses from 25 mm to 50 mm in diameter) usually deliver low magnifications that range between powers of 7 and 15 times. Such low magnifications are unable to show anything but the phases of Venus when it is at its maximum apparent angular diameter, when in its large crescent phases towards the end of an evening apparition or during the start of a morning apparition. Mercury, however, appears as nothing more than a bright pinkish dot, since its apparent diameter is far too small for the planet's phases to be discerned through regular binoculars.

The true field of view (the actual area of sky) taken in by binoculars deceases with magnification. My own 7 × 50 binoculars deliver 7× magnification and a true field of view some 7° wide, allowing frequent observations of lunar, planetary and stellar appulses with both Mercury and Venus. My 15 × 70 binoculars have a true field of view of 4.4°, while my 25 × 100 giant binoculars take in just 3° of the sky.

It will be found that any regular binoculars giving more than 10× magnification must be supported firmly, however lightweight they may be, because higher magnifications exacerbate the slightest movements of the observer's body. However, image stabilized binoculars, first introduced in the early 1990s, eliminate the minor shakes of the user's body. At first glance they resemble regular binoculars of the same aperture but weigh slightly more, and they can be used like normal binoculars. At the push of a button they deliver sharp, vibration-free views courtesy of moving optical elements (most of them require batteries). Stabilized binoculars typically offer quite high magnifications (up to 18×), with apertures from 30 to 50 mm.

15 × 70 binocular observation of the appulse between the very young, very slim waxing crescent Moon (just one day old) and Mercury, viewed on 17 May 2007 by the author. Peter Grego.

Zoom Binoculars

A number of binocular models offer what appears to be an attractive (some might consider highly desirable) zoom facility for general astronomical observation. The magnification of a typical pair of zoom binoculars is capable of being adjusted from, say 15× to as much as 100×. On the face of it, such an instrument would be ideal for the amateur astronomer. It might be imagined that the phases of both Mercury and Venus, plus detail on the Venusian disk, will be readily visible at the higher magnifications delivered by such binoculars. But these instruments have their limitations.

When set at a low magnification, the apparent field of view in zoom binoculars is usually very small – perhaps as small as 40°, far smaller than the apparent field visible in a pair of regular binoculars of comparable magnification. Zooming usually involves physically altering the distance between the lenses inside the eyepiece using a knob or lever, and this is often complicated by the fact that some refocusing is usually necessary after the change in magnification.

Crucially, the alignment of the left and right optical systems in any zoom binoculars needs to be spot-on in order that the brain can produce a merged image from two separate high power images, and this is where most budget zoom binoculars prove wanting.

High magnifications require that the binoculars are mounted on a stand that can be adjusted in altitude an azimuth with as little effort as possible. Even if a well-mounted pair of binoculars does provide good high magnification views of Mercury and Venus, the Earth's rotation makes them appear to move across the field of view at quite a rate of knots. For example, at a magnification of 50×, an apparent field of 50° will equate to a true field of view of one degree, and the planet under scrutiny will move from one edge of the field to the other in around two minutes. This means frequent manual adjustments to the mount if Mercury or Venus is to be kept under scrutiny for any length of time. Driven equatorial binocular mounts are not commonly available, so if a high magnification view of Mercury and Venus is needed, I recommend using a telescope.

Bino-works

A variety of optical configurations are used in binoculars – this is obvious when you look at the vast array of different shapes and sizes available on today's market. As is usually the case, you get what you pay for, but providing binoculars are purchased from a reputable optical dealer, their quality is usually quite good. By shopping with a reputable supplier of astronomical goods you'll have the advantage of pre-purchase advice and post-purchase support.

The quality of the optical system, the optical materials used and the build of binoculars may be noticeable when budget binoculars are compared with high-end binoculars in actual use, although high price does not always guarantee that they are high-end. High-end binoculars use the best optical glass in their object glasses, internal prisms and eyepiece lenses, and they are figured and aligned to exacting standards. Their optical surfaces are usually multi-coated to minimise reflections, and internal baffles block stray light and internal reflections, providing better contrast.

Most binoculars use glass prisms that fold the light between the objective lens and eyepiece lens, and they produce a right way up image, which is of course essential for everyday terrestrial use. 7 × 50 binoculars are ideal general purpose astronomy binoculars. They deliver a wide field of view and have a magnification low enough for the observer to peruse the skies for short periods without requiring the use of a binocular support. 7 × 50 binoculars have an exit pupil of 7 mm. The exit pupil is the diameter of the circle of light that projects from an eyepiece into the eye, and its size can be derived from the aperture of the binoculars divided by their magnification. As the dark adapted eye has an average size of 7 mm, the optimum size of exit pupil for deep dark sky astronomical use measures 7 mm.

There are two basic types of prism binoculars – Porro prism binoculars and Roof prism binoculars. Until a couple of decades ago, most binoculars were of the Porro prism type. Oftenwise, Porro prism binoculars have a distinct 'W' shape, produced by the way that the prisms are arranged to fold the light from the widely separated objective lenses to the eyepieces. American-style Porro prism binoculars are a sturdy design, with prisms mounted on a shelf inside a single moulded case. German-style Porro prism binoculars feature objective housings that screw into the main body containing the prisms – their modular nature makes them more susceptible de-collimation after knocks.

In the past decade, a new style of small binoculars has appeared that use an inverted Porro prism design, leading to a 'U' shaped shell with objective lenses that may be closer together than the eye lenses. The majority of small binoculars produced today are of the Roof prism design, popular because they are compact and lightweight. Roof prism binoculars often have a distinctive H-shape that looks like two small telescopes place side by side – a casual observer may think this indicates a straight-through optical configuration without any intervening prisms. Because of the way that Roof prism binoculars fold the light, they offer generally less contrasty views than Porro prism binoculars.

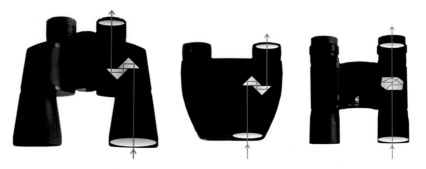

The three main types of binocular – Porro, reversed Porro and Roof prism. Peter Grego.

Telescopes

In addition to seeing detail on the Moon, the dance of Jupiter's four Galilean moons and the rings of Saturn, one of the main reasons people have for buying their first astronomical telescope is to glimpse the phases of Venus. There is no substitute for observing directly through a telescope eyepiece. If asked to choose between having to enjoy Mercury or Venus visually at the telescope eyepiece or from the cosy comfort their home watching a live image from a CCD camera, I'd bet that the majority of amateur astronomers would choose to be out in the field with the telescope, allowing their retinas to be bombarded by photons that have arrived directly from the inner Solar System.

Choosing the right telescope can be a difficult task for a novice. What may be a good telescope for low power deep sky observation may not be the best one for observing the Moon and planets. An astounding array of telescopes of various kinds are advertised in astronomy magazines and on the Internet. Thankfully, these days the optical and build quality of most of these telescopes, an ever increasing proportion of which are Chinese imports, is quite acceptable to all bar the optical puritans.

Regardless of a telescope's aperture or its physical size, the most important thing to be aware of when purchasing any telescope is its optical quality. I would not advise buying a new telescope from any source other than a reputable telescope dealer. Sensational newspaper advertisements and department store bargains ought to be avoided – they won't normally allow you the opportunity to inspect the instrument prior to purchase or use it on a trial-return basis. The ultimate test of a retailer's confidence in the quality of their goods and a measure of their customer friendliness is that they welcome an examination of the goods. Any major flaws or in the optics or dings and defects on the exterior of the instrument are quickly revealed under the bright lights of the store, and an observational test will bring to light any optical problems.

Sadly, newspaper advertisements announcing massive stock clearances of telescopes and binoculars are often full of outrageous hyperbole, with claims of extravagant magnifications and the ability of the instrument to show all the wonders of the Universe in glorious detail. These advertisements attempt to blind the unsuspecting novice with 'science' in an attempt to hide the fact that their instruments may be made entirely of plastic – even the lenses. Such optical monstrosities are utterly useless for any kind of observing, and the horribly distorted images that they deliver can put the novice off astronomy altogether.

Any reputable company that specialises in selling and/or manufacturing optical instruments will give by far the best deal, in terms of price, service and advice. All the major astronomical equipment retailers advertise in astronomy magazines, and most of them produce a product catalogue that can be browsed online or in printed form.

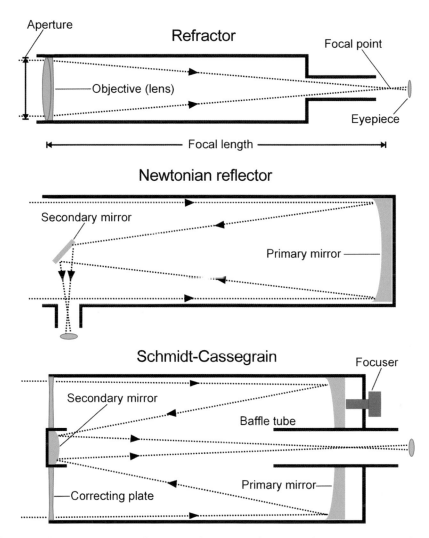

Illustrating the three main type of amateur telescope – achromatic refractor, Newtonian reflector and Schmidt-Cassegrain. Peter Grego.

Refractors

If asked to imagine a typical amateur astronomer's telescope, an image of a refractor will jump into most people's minds. Refractors consist of a closed tube with an objective lens at one end and an eyepiece at the other end. Light is collected and focused by the objective lens (the light is bent, or refracted, hence the name 'refractor') and the eyepiece magnifies the focused image. A telescope's focal length is the distance between the lens and the focal point, usually expressed as a multiple of the objective lens diameter or in millimetres. So, a 100 mm f/10 lens has a focal length of 1000 mm, and a 150 mm f/8 lens has a focal length of 1,200 mm. Eyepieces also have a focal length, but this is always given in millimetres. The smaller an eyepiece's focal length, the higher the magnification it will deliver. The magnification of any particular eyepiece/telescope combination can be calculated by dividing the telescope's focal length by the eyepiece's focal length. For example, a 12 mm eyepiece used on a telescope with a focal length of 1500 mm will give a magnification of 125×; the same eyepiece used on a telescope of 900 mm focal length will deliver a magnification of 75.

Galilean telescopes are the simplest form of refractor, with a single objective lens and a single eyepiece lens. They suffer greatly from chromatic aberration, caused by the splitting of light into its component colours after it is refracted through glass, and spherical aberration, caused by light not being brought to a single focus. Through a Galilean telescope a bright object like Venus appears surrounded by fringes of vividly coloured light, and the whole image appears washed out and blurry. Cheap small telescopes attempt to alleviate the worst effects of aberration by having large stops placed inside the telescope tube, preventing the outer parts of the cone of light from travelling down to the eyepiece. This crude trick simply makes a poor image appear slightly less poor, and the presence of the stop reduces the aperture of the instrument, reducing its light grasp and resolving power.

Small though they may be, good quality astronomical finderscopes should not be confused with Galilean telescopes. Finderscopes are low power refractors that are precisely aligned with the larger telescopes that they are attached to in order that the observer can quickly and accurately locate bright celestial objects. When centred on the finder's cross hairs, the object is also visible at a higher magnification in the main instrument. Finderscopes have achromatic objective lenses (typically 20–50 mm) and have fixed eyepieces that can be adjusted for focus. Straight through finderscopes deliver an inverted view, so they are unsuitable for terrestrial use.

Small Refractors

A well-made telescope, of any size, is capable of providing a pleasing view of Mercury and Venus. Under poor seeing conditions, when the atmosphere is shimmering and the stars scintillate wildly, attempts to observe at high magnifications through large instruments can prove futile, as the planets can take on the appearance of wobbling blobs of jelly. On these nights a small telescope will sometimes deliver an apparently sharper, more stable image than the large telescope, since a small telescope will not resolve as much atmospheric turbulence as a larger one.

Small telescopes have a number of other advantages. Being lightweight, they are eminently portable, and they can be carried around an observing site to avoid local obstacles to the sky, such as trees and buildings. A small, relatively inexpensive telescope might be considered expendable, and for this reason the observer may actually be inclined to use it more often than a 'precious' high-end telescope – after all, accidental damage to the external structure or the optics of a cheap telescope is not nearly as soul-destroying as bashing an instrument that cost ten times as much.

The least expensive small telescopes, including those hand-held old style repro-
duction brass 'naval' telescopes using draw tubes to focus the image, usually have a
fixed eyepiece that delivers a constant magnification. Some fixed eyepiece telescopes
of this nature are a little more sophisticated by allowing some variation of magnification; for occasional Mercury and Venus spotting I use a very nice one with a
30 mm lens whose magnification is adjustable between 10–30× (with no variation
in the apparent field of view), some 140 mm long when folded and 340 mm when
extended.

Two handy little telescopes
a brass drawtube 30
mm refractor with a 10–30
variable magnification,
compared with a 10× 25
monocular. Peter Grego.

Somewhat more versatile is a telescope that allows the eyepieces to be inter-
changed, allowing the magnification to be alternated between low to high powers.
Two or three eyepieces, along with perhaps a magnifying eyepiece called a Barlow
lens, are often provided – these are usually small plastic mounted eyepieces with
a 0.965-inch barrel diameter, often of a very basic optical design. These eyepieces
may deliver poor quality views with incredibly narrow apparent fields of 30° or
even smaller.

Eyepieces ought never be considered accessories – they are as vital to the performance of a telescope as its objective lens or mirror. So, if a small instrument is
not performing as well as expected, don't shove it in the cupboard straight away –
replace the eyepieces with some better quality ones, purchased from an optical
retailer. The most widely available eyepieces have a barrel diameter of 1.25-inches,
and these can be mounted to a .965-inch eyepiece tube using an adapter or a
hybrid star diagonal (a mirror that flips the image 90°). Plössl eyepieces deliver
an apparent field of about 50°, and quality budget versions of this eyepiece design
are available. Good quality eyepieces can transform a small budget telescope into a
good performer at low to medium magnifications (see below for more information
about eyepieces).

A Better Class of Glass

Good quality astronomical refractors have an achromatic objective lens that
comprises two specially shaped lenses of different types of glass, nestled closely

Here Jacy Grego observes through a 60 mm refractor (800 mm focal length) and a quality 16 mm Zeiss 0.965-inch orthoscopic eyepiece, giving a comfortable magnification of 50× to view Venus at its larger crescent phases. Peter Grego

together. These lenses attempt to refract all the different wavelengths of light to a focus at a single point. Achromatic objectives do not eliminate the false colours of chromatic aberration altogether, but generally speaking the effects are less noticeable in longer focal length refractors. Many budget imported achromatic refractors have focal lengths ranging between f/8 to as short as f/5, and although these instruments do display a degree of chromatic aberration, mainly in the form of a violet fringe around bright objects such as Venus, they offer good resolution and a reasonable amount of contrast.

Chromatic aberration in budget achromatic refractors can be reduced with a minus-violet contrast boosting filter that screws into the eyepiece. Alternatively, chromatic aberration can be physically minimised using a specially designed lens (one brand is called a 'Chromacorr') attached to the eyepiece; the lens takes all the wavelengths of light and refocuses them to as near a single point as possible, potentially transforming a budget achromatic refractor into a telescope that approaches the performance of a high-end apochromatic refractor.

Apochromatic refractors use special glass in their two or three element objective lenses to bring the light to a pin-sharp focus, delivering images that are virtually free from the effects of chromatic aberration. Planetary views through an apochromat are almost completely free of aberration and are of high contrast, rivalling the kind of view to be had through a good quality long focal length Newtonian reflector (see below). Aperture for aperture, an apochromatic refractor will, as a general rule, work out at more than ten times as expensive than a budget achromatic refractor.

Refractor Care

Refractors are ready to use straight out of the box and usually require little maintenance. Their objective lenses are aligned in the factory and sealed in their cell. If a refractor is working well there is no reason to unscrew an objective lens and remove it from its cell – although being naturally curious, a great many amateur astronomers have been tempted to do it just to discover how the thing is put together. It is not recommended, as the reassembled components will more than likely perform less well than before. For this reason, small specks of dust in between the components should be tolerated rather than removed.

Over time, the external surfaces of lenses can accumulate a fair amount of dust and debris, but great care must be taken when cleaning them. Many lenses have an anti-reflection coating, a thin layer which can be disturbed if the lens is not cleaned in the proper manner. A lens of any sort should never be rubbed vigorously with a cloth; any bits of dirt might scratch the coating or the lens itself. Dust particles should be carefully removed with a soft optical brush or an air puffer, and any residual grime can be gently removed with optical lens wipes, each used once and applied in a single stroke. Condensation on a lens should be allowed to dry naturally, and never rubbed off.

Reflectors

As their name suggests, reflecting telescopes use mirrors to collect and focus light. They have a specially shaped concave primary mirror to collect light and reflect it to a sharp focus, free from the effects of chromatic aberration which dog refractors. Reflectors (especially ones of short focal length) are prone to a phenomenon called spherical aberration, where the margins of an image can appear fuzzy and out of focus; this problem can be addressed overcome by using mirrors with accurately figured paraboloidal surfaces.

Newtonian reflectors are the most popular reflecting telescope design. They use a concave primary mirror held in an adjustable cell at the bottom of the tube and a smaller flat secondary mirror mounted on a 'spider' near the top of the tube that reflects light sideways at a 90° angle through the tube into the eyepiece. A well-collimated long focal length (f/10 and longer) Newtonian will deliver superbly detailed high magnification planetary views.

Cassegrain reflectors have a primary mirror with a central hole. The primary reflects light onto a small convex secondary mirror, which reflects the light back down the tube, through the hole in the primary mirror and into the eyepiece. Prone to the optical aberrations of astigmatism and field curvature, most Cassegrains are large observatory-sized instruments with focal lengths ranging from f/15 to f/25 – excellent for planetary studies at high magnification.

Reflector Collimation and Care

Reflectors require much more care and attention than refractors. A sudden knock to the tube can cause misalignment of the primary mirror in its cell or the secondary in its spider, leading to poor quality images, including dimming, blurring and multiple images near the point of focus. Even a brand new Newtonian taken out of its carton is likely to require recollimation in order to align the optical components as precisely as possible.

Newtonian primary mirror adjustment is usually obtained by means of knobs or wingnuts at the base of the mirror cell, but the secondary's alignment will usually require a small screwdriver or Allen key. Collimation can be time consuming and a little tricky for novices, but new telescopes should be provided with adequate instructions. In any case, there are many Internet resources that explain the process in detail. There are ways to achieve good collimation, including laser collimators and a device called a Cheshire eyepiece.

Most reflectors are not sealed when in use, and many have open-tube designs that allow airborne dirt to settle onto the primary mirror surface. Mirrors are coated with a wafer thin layer of reflective aluminium, and special coatings can extend the life of a mirror by a factor of two or three. It's best to allow a mirror to dry off naturally after an observing session, and then protect the mirror from exposure to the open air when not in use. However, all mirrors accumulate a layer of dust and bits of debris over time, and a primary mirror can look disconcertingly filthy when illuminated by a flashlight at night. Debris on the mirror scatters light, and as the mirror gets grubbier it becomes less effective, producing a decline in image contrast.

An aluminised mirror must be cleaned very carefully, since hard bits of debris scraped across the thinly aluminised surface will leave scratches like skates on ice. Loose debris can be blown away with a puffer or a canister of compressed air, and the mirror can be cleaned with cotton wool and lens cleaning fluid or lens wipes – this must be performed very gently, with a single stroke per cleaning wipe, and is best done with the mirror taken out of its cell.

One way of extending the life cycle of a reflector is to stretch a piece of optically transparent film over the telescope aperture in order to seal the top of the tube (the bottom of a Newtonian is often open, allowing the free circulation of air for better image quality). This see-through material, such as Baader TurboFilm, is available in large sheets that can be cut to fit small to medium-0sized reflectors. For the best image quality, the sheet should ideally be evenly taut, without wrinkles. When the film itself gets dirty, another disk can easily be produced. A well protected, well cared for Newtonian mirror can last more than a decade before requiring a fresh re-aluminising.

Catadioptrics

Catadioptric telescopes use a combination of mirrors and lenses to collect and focus light. There are two popular forms of catadioptric – the Schmidt-Cassegrain telescope (SCT) and the Maksutov-Cassegrain telescope (MCT).

In the SCT, light enters the top of the telescope tube through a sheet of flat-looking glass with a large secondary mirror mounted at its centre. This glass sheet is actually a vital optical element called a corrector plate; it is specially shaped to refract the light onto an internal primary mirror (usually with a focal length of f/8). The light is then reflected back up the tube to the convex secondary mirror, which in turn reflects it back down the tube and through a central hole in the primary mirror into the eyepiece.

The relatively large size of the secondary mirror in a SCT produces a degree of diffraction, or 'fuzzing' of light, that can slightly affect image contrast. An optically good, well-collimated SCT will deliver superb views of the planets. Moreover, due to their design, a number of useful accessories can be attached to the 'visual back' (the part of the telescope that the eyepiece normally fits into) – accessories useful to the planetary imager and observer include filter wheels, SLR cameras, digicams, camcorders, webcams and CCD cameras.

MCTs use a spherical primary mirror and a deeply curved spherical lens (a meniscus) at the front of the tube. The secondary mirror in the MCT is actually formed by a small spot aluminised directly on the interior surface of the meniscus. Light enters the tube through the meniscus, refracts onto the primary mirror and is reflected into the eyepiece via the secondary mirror and a central hole in the primary.

Although MCTs superficially resemble SCTs, they tend to be far better performers on the Moon and planets. With their long focal lengths and excellent correction for spherical aberration, MCTs deliver excellent resolution, high contrast views.

The author with his 200 mm SCT and 127 mm MCT. Peter Grego.

Telescopic Resolution

On nights of really good seeing, the resolving power (R, in arcseconds) of a telescope of aperture diameter (D, in millimetres) can be calculated using the formula R = 115/D.

In terms of observing Mercury and Venus, note that Mercury's minimum apparent angular size is 4.8 arcseconds, while that of Venus is 10 arcseconds. At maximum, Mercury can be as large as 10 arcseconds across, while Venus can exceed one arcminute. Mercury's average diameter at maximum elongation is 7.5 arcseconds, while that of Venus is 25 arcseconds.

Aperture (mm)	Resolution (arcsec)	Suggested max magnification
30	3.8	60
40	2.9	80
50	2.3	100
60	1.9	120
80	1.4	160
100	1.2	200
150	0.8	300
200	0.6	400
250	0.5	500
300	0.4	600

Eyepieces

Even if a telescope has perfectly figured lenses or mirrors, it will not perform at its best if a poor quality eyepiece is used. An eyepiece's magnification can be calculated by dividing the telescope's focal length by the focal length of the eyepiece. A 20 mm eyepiece used on a telescope with a focal length of 1,500 mm will deliver a magnification of 75 (1,500/20 = 75). The same eyepiece will magnify ×40 on a telescope with a focal length of 800 mm (800/20 = 40).

For planetary observation, it is recommended to have at least three good quality eyepieces that deliver low, medium and high magnifications. A low power eyepiece is excellent to follow Mercury or Venus during an appulse with the Moon, other planets or stars. Using a telescope of 1,000 mm focal length, a 20 mm eyepiece with a 50° apparent field will show an actual field of around one degree, delivering a magnification of ×50. A high power eyepiece is considered to be one that delivers a magnification double the telescope's aperture in millimetres – for example, ×200 on a 100 mm refractor. High powered scrutiny of Mercury or Venus can only be performed when seeing conditions allow.

Eye relief is the maximum distance from the eyepiece that the eye can comfortably be positioned to see the full field of view. It can be gauged by holding an eyepiece up to the light and determining how far away your eye is from the front lens for you to see the entire circular field stop outlining the field of view. Spectacle wearers find that eyepieces with a minimum of 15 mm eye relief allow comfortable viewing while keeping the spectacles on, obviating the need to remove them in order to get the eyeball up close to the eye lens. Some eyepiece designs have better eye relief than others; usually, shorter focal length eyepieces have less eye relief. Spectacle wearers can get around this limitation and obtain high magnification views using a Barlow lens in combination with a longer focal length eyepiece with good eye relief.

Astronomical eyepieces are produced in three barrel diameters – 0.965-inch, 1.25-inch and 2-inch. Tiny 0.965-inch eyepieces come with many budget small telescopes; they are usually made of plastic, of a very unsophisticated optical design and are of poor quality. Good quality 0.965-inch eyepieces are hard to find these days, so it's better to upgrade to 1.25-inch eyepieces, which are the most common diameter available. Most telescope focusers are built to accept 1.25-inch eyepieces, and many of them can accommodate 2-inch barrel eyepieces as well. 2-inch barrel eyepieces can be hefty beasts with incredibly large lenses. They usually accommodate very wide angle, long focal length optical systems that are ideal for wide angle, low to medium magnification astronomical observation.

Budget telescopes are usually supplied with Huygenian, Ramsden or Kellner type eyepieces which all have very restricted apparent fields of view – more a deep sea diving experience than a spacewalk! Huygenian, Ramsden and Kellner eyepieces are unsuitable for high magnification planetary observing.

The Huygenian is a very old design, consisting of two plano-convex lenses; the convex sides both face the incoming light, and the focal plane lies between the two lenses. Huygenians are undercorrected (the rays from the outside zone of the lenses come to a shorter focus than rays focused by the central parts), but the aberrations from each lens effectively cancel each other out. Huygenians deliver very small apparent fields of 30° (or even smaller), and are only suitable for use with telescopes with a focal length of f/10 or greater. Huygenians deliver poor eye relief.

Ramsdens are another very old design, like Huygenians consisting of two plano-convex lenses, but both convex sides face each other (sometimes the lenses may be cemented together to provide better correction); the focal plane lies in front of the field lens (the lens that first intercepts the light). Ramsdens deliver flatter, slightly larger apparent fields of view than Huygenians, but they are prone to a greater degree of chromatic aberration, are poor performers on short focal length telescopes and offer poor eye relief.

Kellners are the least ancient of the three basic designs. Similar to the Ramsden, their eye lens (the lens closest the eye) consists of an achromatic doublet. Kellners deliver better contrast views than Huygenians or Kellners, with fields of about 40°, but annoying internal ghosting is invariably seen when viewing bright objects like Venus. Like the Ramsden, the Kellner's focal plane is located just in front of the field lens, so any minute particles of dust that happen to land on the field lens will be seen as dark silhouettes against any bright field. Kellners with focal lengths longer than 15 mm perform the best, while shorter focal lengths can produce a blurred effect around the edge of the field along with chromatic aberration. Kellners have good eye relief.

Monocentric eyepieces consist of a meniscus lens cemented to either side of a biconvex lens. Despite having narrow apparent fields of view of around 30°, monocentrics deliver excellent crisp, colour-free, high contrast images of the planets, completely free from ghost images, and they can be used with low focal length telescopes.

Orthoscopic eyepieces comprise four elements – an achromatic doublet eye lens and a cemented triplet field lens. They produce a flat, aberration-free field, and deliver very good high contrast views of the planets. Their apparent field of view varies from around 30° to 50° and they have good eye relief.

Erfle eyepieces have multiple lenses (usually a set of two achromatic doublets and a single lens, or three achromatic doublets) that deliver a wide 70° field of view with good colour correction. Erfles perform at their best when used with long focal length telescopes, and the best versions are of 25 mm focal length and greater. However, the definition at the edge of the field of view tends to suffer, and their multiple lenses produce internal reflections and annoying ghost images when bright objects are viewed.

Today's most popular eyepiece is the Plössl, a four element design that produces good colour correction and an apparent field of view around 50° that is flat and sharp up to the edge of the field. They can be used with telescopes of very short focal length. Standard Plössls with long focal lengths have a good degree of eye relief. Lower focal length Plössls of a standard design have poor eye relief, so they

Observing Mercury and Venus

may be a little awkward to use for high magnification views of the planets, but versions are available with longer eye relief.

Modern demand for quality wide field eyepieces has led to the development of designs such as the Meade UWAs, Celestron Axioms, Vixen Lanthanum Superwides and Tele-Vue Radians, Panoptics and Naglers. These all deliver excellently-corrected images with wonderfully large apparent fields of view, and all have good eye relief. With their 80° plus apparent fields, Naglers are awesome eyepieces – and, like all quality super wide field eyepieces, they retail at an awesome price. A set of four Nagler eyepieces can cost more than a brand new 200 mm SCT. Some of the older style long focal length Naglers are extremely big and heavy, and switching eyepieces requires rebalancing the telescope.

Zoom eyepieces remove the necessity of changing between eyepieces of various focal lengths to vary the magnification. A number of reputable companies, including Tele-Vue, sell premium zoom eyepieces. Zoom eyepieces have been around for many years, but they have not yet achieved widespread popularity among serious amateur astronomers, perhaps because zooming is perceived to be a novelty associated with many budget binoculars and telescopes. Good quality zoom eyepieces are by no means frivolous items. They achieve a range of focal lengths by adjusting the distance between some of the lenses. A popular premium 8–24 mm focal length zoom eyepiece has a narrow 40° apparent field when set at its longest focal length of 24 mm, but as the focal length is reduced, the apparent field enlarges, up to 60° at 8 mm. A good zoom eyepiece can replace a number of regular eyepieces, and at a fraction of the cost.

Binocular Viewers

Viewing with two eyes, rather than one, has distinct advantages. It is more comfortable to use both eyes, and the view is more aesthetically pleasing. With two eyes, a two-dimensional image takes on a near three-dimensional appearance, and it is often found that finer detail can be discerned.

Binocular viewers split the beam of light from the telescope's objective into two components which are reflected into two identical eyepieces. Most binocular viewers require a long light path, and they will only work on an instrument whose focuser can be racked in enough for the prime focus to pass through the convoluted optical system of the binocular viewer. Refractors and catadioptrics (SCTs and MCTs) are suitable for use with binocular viewers, but they may not be able to come to a focus through a standard Newtonian telescope. Binocular viewers are designed to be used with two identical brand eyepieces of the same focal length. Eyepieces of 25 mm focal length and shorter are recommended, since vignetting around the edge of the field becomes apparent when much longer focal length eyepieces are used. The use of two premium zoom eyepieces will save having to swap eyepieces to alter the magnification.

Scopes and Seeing

Here on the Earth's surface, we view space through a thick layer of atmosphere – 99 percent of the Earth's atmosphere lies in a layer just 31 km deep. Most of the problems are caused by the bottom 15 km of the atmosphere.

Clouds are the most obvious impediment to astronomical observation, but even a perfectly cloud-free sky can be useless for telescopic observation. The atmosphere is full of air cells of varying sizes (2–20 cm) and density, and light is refracted slightly differently as it passes through each cell. The worst seeing is produced when the air cells are mixing vigorously, making the light from a celestial object appear to jump around. The level of turbulence observed also varies with the height of the observed object above the horizon – a telescope that is pointed low down will be looking through a far thicker slice of air than one pointed high. The observer's immediate environment also plays a significant role in how good the image is. A telescope brought out into the field needs some time to cool down. Chimneys, houses and factory roofs that give off heat produce columns of warm air that mix with the cold night air to warp the image.

Seeing varies from 0.5 arcsecond resolution on an excellent night at a world-class observatory site, to 10 arcseconds on the worst nights. On nights of poor seeing it's hardly worth observing the planets with anything but the lowest powers, since turbulence in the Earth's atmosphere will make the a planetary image appear to roll and shimmer, rendering any fine detail impossible to discern. For most of us, seeing rarely allows us to resolve planetary detail finer than 1 arcsecond, regardless of the size of the telescope used, and more often than not a 150 mm telescope will show as much detail as a 300 mm telescope, which has a light gathering area four times as great. It is only on nights of really good seeing that the benefits of the resolving power of large telescopes can be experienced – such conditions sadly occur all too infrequently for most amateur astronomers.

Telescope Mounts

It is important that a telescope is attached to a sturdy mount, and that the telescope is able to be moved without any difficulty in order to keep the planets in the eyepiece as the Earth rotates. The simplest form of mount is a telescope inserted into a large ball that can freely and smoothly rotate in a cradle – several small budget reflectors are mounted in such a manner, and they are great fun to use, although they are unable to be operated in anything other than a manual fashion.

Altazimuth Mounts

Altazimuth mounts enable the telescope to be moved up and down (in altitude) and from side to side (in azimuth). Small undriven table-top altazimuth mounts are often provided with small refractors, but the quality of their construction can be poor. Most problems are caused by inadequate bearings on the altitude and azimuth axes – they may be too small and the right amount of friction may be difficult to achieve. Overly tight bearings will result in too much force being used to overcome the friction, and smooth tracking cannot be achieved. Better altazimuth models are provided with slow motion knobs that allow the telescope to be moved without having to push the telescope tube around. If the mount itself is lightweight and shaky then it is liable to be buffeted by the slightest wind, rendering it unable to be used in the field – it may be better to attach the telescope to a good quality camera tripod.

Dobsonians are altazimuth mounts that are used almost exclusively for Newtonian reflectors of short focal length. Since their invention several decades ago, they have become highly popular because they are simple to build and easy to use. Dobsonians consist of a ground box with an azimuth bearing and another box that holds the telescope tube. The altitude bearing is at the centre of balance of the telescope tube, and it slips neatly into a recess in the ground box. Low friction materials like polythene, Teflon, Formica and Ebony Star are used for the load-bearing surfaces, enabling the largest Dobsonians to be moved around at the touch of a fingertip. Lightweight structural materials such as MDF and plywood make Dobsonians strong but highly portable, and commercially produced Dobsonians range from 100 mm to half-metre aperture Newtonians.

It is not too difficult a task to keep a planet in the field of view of a telescope mounted on an undriven altazimuth or Dobsonian mount up to magnifications of ×50. The higher the magnification, the faster the planet under scrutiny appears to move across the field of view, and more frequent small adjustments need to be made to keep it centred in the field. If the observer wants to make an observational sketch, the limit of magnification for an undriven telescope is ×100 – anything higher and the instrument will need adjusting after each time the drawing is attended to – a tedious process that will double the length of time that a drawing should take to complete. At ×100, a planet that is centred in the field will take about 30 seconds to move to the edge of the field. High magnification with undriven telescopes also demands a sturdy mount that does not shake unduly when pushed, and smooth bearings that respond to light touch and produce little backlash – qualities found in only the best altazimuth and Dobsonian mountings.

Equatorial Mounts

Serious planetary observation requires a telescope to be mounted on a sturdy platform with one axis parallel to the rotational axis of the Earth, the other axis at right angles to it. In an undriven equatorial telescope a planet can be centred in the field of view and kept there with either an occasional touch on the tube or the turn of a slow motion control knob that will alter the pointing of on one axis – far easier than having to adjust the telescope on both axes of an altazimuth mounted telescope to keep a celestial object within the field of view. A properly aligned, well-balanced driven equatorial allows the observer more time to enjoy observing without worrying about it quickly drifting out of the field of view. Equatorial clock drives run at a 'sidereal' rate, enabling celestial objects that are centred in the field to remain there over extended periods of time, depending on how well the equatorial's polar axis is aligned, the accuracy of the drive rate and the apparent motion of the celestial object.

German equatorial mounts on aluminium tripods are most often used with medium to large refractors and reflectors. A telescope mounted on a German equatorial is able to be turned to any part of the sky, including the celestial pole. Schmidt-Cassegrain telescopes are commonly fixed to a heavy duty fork type equatorial mount. The telescope is slung between the arms of the fork, and the base is tilted to point to the celestial pole. When their visual back has a particularly large accessory attached, say a CCD camera, these instruments are sometimes unable to view a small region around the celestial pole because the telescope cannot swing fully between the fork and the base of the mount – this is no problem for observers of Mercury or Venus because neither planet ever gets near the celestial pole.

Many amateurs choose to keep their mount and telescope in a shed and set it up each time there is a clear night – setting up requires some time, and is usually done in several stages. The mount's polar axis must be at least roughly aligned with the celestial north pole for it to track with any degree of accuracy. A tripod can be difficult to adjust on sloping ground, and the legs not only pose something of a navigational hazard when walking around the instrument in the dark, but the seated observer will invariably knock the tripod from time to time, producing vibration of the image. To eliminate the time-consuming chore of setting up and polar aligning each observing session, some amateurs construct a permanent pier, set in concrete, upon which their German equatorial mount can be fixed and aligned, or to which their entire SCT and mount can be quickly and securely fixed.

Computerised Mounts

Computers are revolutionising amateur astronomy in many ways, and one of the most visible signs of this is the increasing preponderance of computer controlled telescopes. They come in all varieties – small refractors mounted on computer-driven altazimuth mounts to large SCTs on computerised fork mounts and German equatorials. Some standard undriven equatorial mounts can be upgraded to accept either standard clock drives or computerised drives. Once details of the observing location and the exact time are input, a computerised telescope can automatically slew to the position of any celestial object above the horizon at the touch of a few buttons on a keypad.

Smaller computerised telescopes tend to have fairly insubstantial mounts that are incapable of supporting much more weight than the telescope itself while maintaining good pointing and tracking accuracy. While such instruments are acceptable for visual planetary observation, they may not withstand the addition of a heavy accessory such as a digital camera or binocular viewer. Larger computerised telescopes of the SCT varieties produced by Meade and Celestron, for example, are constructed well enough to accommodate hefty accessories.

A computerised telescope is capable of automatically slewing to the position of Mercury or Venus and track it accurately at the touch of a button, and basic information about these planets can be displayed on the keypad's viewscreen. It is a great bonus to be able to locate Mercury and Venus with relative ease during the daytime using a computerised telescope, since the alternative methods of locating them – low magnification sweeps and using setting circles – can prove difficult and time consuming, even if the instrument is accurately polar aligned. Besides, the setting circles on all but the best astronomical mounts are usually not accurate enough to be up to the task, and for all intents and purposes may be considered as decorative, rather than practical features on budget instruments.

Chapter 5

Recording Mercury and Venus

Recording Techniques

Traditional photography has never been able to show as much detail on Mercury and Venus as a drawing made at the telescope eyepiece under identical conditions. Until CCD imaging became a powerful tool in amateur astronomers' hands in the last decade of the 20th Century, drawing at the telescope eyepiece was the only means of recording the subtle shadings on the surface of Mercury and the cloud patterns and other phenomena exhibited by Venus.

Despite the fact that the markings of Mercury are notoriously understudied, and the fact that Venus' cloud patterns are continuously changing, often in surprising ways, few professional observatories ever turn their big telescopes towards these two planets – other than perhaps to impress visitors and students or to test out new equipment. Of course, many amateur astronomers never consider undertaking systematic studies of the inferior planets, but instead choose occasionally to view Mercury and Venus for the sheer challenge and the visual pleasure that they provide.

Imaging and observational drawing of Mercury and Venus have been practices entirely undertaken by amateur astronomers for many decades. Happily, these planets offer plenty of interest to occupy the observer, and indeed, the dedicated amateur can expect to discover transient Venusian atmospheric phenomena.

Mercury is a difficult object to observe at those times during the post-dusk or pre-dawn twilight when it is able to be easily located and centred in the telescopic field of view. There exist only narrow observing windows of a few weeks each elongation, in addition to the planet's low altitude at dawn or dusk, its small apparent diameter and its proximity to the Sun (never venturing further than 28° in elongation). It's fair to say that few amateur astronomers have obtained a really good look at Mercury, nor do many regard it as much more than a tiny, shimmering, featureless object – nice to see, but difficult to truly appreciate. This is a great pity, because the planet

For different reasons Venus is a somewhat neglected planet, considered by some cursory telescopic viewers to be pretty – but pretty bland. Venus is on view for many months during each elongation in the morning or evening skies, often set brilliantly against a dark sky background beyond astronomical twilight. However, it very presence over extended periods during each year, and its persistent brilliance, may be reasons why many amateur astronomers grow reluctant to view it telescopically on a regular basis – a case of familiarity breeding complacency rather than contempt.

Venus is so bright – indeed, dazzlingly so through the telescope eyepiece – that many observers find it difficult to discern the planet's phase at low magnifications (particularly when a small gibbous disk), nor can many cursory observers discern its cloud features with ease. Venus' cloud features are readily visible in ultraviolet (UV) wavelengths, and some people are more sensitive to UV light than others. Those lacking in visual UV sensitivity used to be those most vociferous in casting doubt upon observations of Venus' cloud features.

Observational Drawings

Unlike many other branches of Solar System astronomy, such as imaging detail on the Moon, Jupiter and Saturn, the competent visual observer can still produce observations that show as much detail as the best CCD images. Visual observers can make the most out of mediocre seeing conditions, teasing out finer planetary detail during those moments of improved seeing.

Those making observational drawings should attempt to depict the features visible on Mercury and Venus as accurately and as unambiguously as possible. The drawings need not be great masterpieces of art, and no observer's work need be considered of more value than another's because it simply looks nicer. The aim is to attend to the fine detail visible through the eyepiece and record it to the best of your abilities. The finished results will be the products of your efforts and a permanent record of your planetary forays. Don't throw them away – keep all your observations in a folder and you may be pleasantly surprised at how your skills of planetary depiction improve over time.

Outline Blanks

Lack of preparation can often lead to disappointing results. It's not uncommon to see the would-be planetary observer fumbling around at the eyepiece – pencil in one hand, sketchpad in the other, flashlight in mouth – attempting to depict the outline of a planet by freehand or by drawing around the edge of an eyepiece barrel. I've done it myself many times in the past, and needless to say, those particular attempts at planetary observation are kept well away from the scrutiny of others.

So, before observing Mercury or Venus, prepare a circular outline blank – one of 50 mm diameter is recommended. If intensity estimates (see below) are to be made in addition to a pencil sketch, two equal-sized circles should be drawn side by side on the same piece of paper; this will minimize having to write down identical details on two sheets, and it will avoid confusion if the sketch and intensity estimate become separated. Alternatively, planetary blanks and observing forms can be downloaded and printed off directly from various Internet resources:

SPA Planetary Section: http://www.popastro.com/sections/planet/forms.htm
ALPO Mercury Section: http://www.lpl.arizona.edu/~rhill/alpo/mercstuff/
    mercfrm.jpg
ALPO Venus Section: http://www.lpl.arizona.edu/~rhill/alpo/venustuff/venus1.pdf
BAA Mercury and Venus Section: http://www.take27.co.uk/BAA_MV/
    BAA_MVS.html (observing blanks available from the Director).

When printing off report forms I use a monochrome laser printer and 100 gsm white paper, because inkjet prints have an annoying habit of smudging once they become damp. Now, to increase the accuracy of an observational drawing of Mercury or Venus it is perfectly acceptable to include the predicted phase on the outline blank, prior to observing. This may be calculated using tables published in astronomical ephemeredes and drawn manually, or the phase can be drawn accurately using any one of a number of astronomical programs. Now, the predicted phase of Venus does not always match the observed phase, and if the pre-printed outline blank is black-bordered and phase corrected it will not be possible to alter the phase according to your own observations; nor will it be possible to depict Venusian phenomena such as the ashen light, cusp extensions, and protrusions at the limb and terminator, if observed, on the outline blank itself. In addition, intensity estimates are hampered by using a black-bordered blank, since the estimates must be written upon the drawing itself rather than indicated by lines pointing to features on the drawing. My own preference is to use a simple circular outline blank with a pre-drawn curve for the terminator, for both sketch and intensity estimate, enabling the most information to be portrayed in as accurate and uncluttered a fashion as possible.

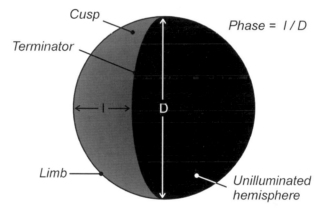

Showing the main features of a crescent phase. Peter Grego.

Pencil Sketching Technique

A set of soft-leaded pencils, from HB to 5B is recommended. Outlines of the most prominent planetary features should initially be drawn very lightly, using a soft pencil, giving you the chance to erase anything if the need arises. It is best not to depict the dark sky around a planet. You're unlikely to see any really dark features, but any unusually dark areas should be drawn by applying minimal pressure on the paper using layers of soft pencil rather than a single layer made with heavy pencil pressure. Soft graphite smudges really well, and good use of smudging along the terminator or along the edges of cloud features can produce smooth blends.

It is reasonable to set yourself about half an hour per observational drawing. Patience is essential – even if the clouds are threatening to obscure the planet from view, or if your fingers are feeling numb with cold – because a rushed sketch is bound to be less accurate.

Don't expect to see much in the way of detail on Mercury or Venus if you hastily set up your telescope and begin observing straight away. If you're taking your telescope from a warm indoor environment to a colder outdoors, allow your telescope a good while to acclimate and cool down. This is especially important with sealed-tube instruments like SCTs and MCTs, otherwise you'll get a poor image caused by turbulent air currents, and the observation will hardly be worth making. My own 200 mm (8-inch) SCT and 127 mm (5-inch) MCT are left outside in the backyard for around an hour before observing in earnest – if it's an early evening session, I put the instrument in an unheated shed prior to setting it up or set it up in a spot that will remain shaded until sundown.

Digital Drawing

Observational drawings made directly onto the touch-sensitive screen of a PDA (Personal Digital Assistant), UMPC (Ultra Mobile PC) or tablet PC make an excellent alternative to making pencil sketches at the eyepiece.

Having used a number of these small computers over the years, I've found that the older WinCE PDAs, such as the HP Jornada range (with the exception of the keyboard-equipped 680 to 720 models) are quite useful for astronomical sketching – more suited to the purpose, indeed, than a number of the newest Windows Mobile PocketPC devices such as the XDA Orbit which have a much smaller screen size. Screen size is often smaller on many Palm powered PDAs, too, owing to the presence of a graffiti pad (for handwritten input and character recognition) beneath the screen. For planetary sketching my currently preferred PDA is an PocketPC powered SPV M2000.

There are numerous capable graphics programs for PDAs available to be downloaded, some of which are freeware, shareware or trialware commercial programs, allowing you to try before you buy. For my SPV M2000 I prefer to use the simple, easy to use and versatile Mobile Atelier by NeFa Studio; freeware versions of Mobile Atelier for use with older WinCE devices are also available.

I often use a drawing program (Corel Paint) or on my tablet PC for lunar and planetary observation during those evenings when I'm running astronomical software in the field or performing CCD astronomy through my tablet PC. In inclement conditions I use a Hammerhead HH3, an ultra-tough tablet PC that can withstand fair degrees of cold and heat, moisture, being trodden on and being dropped to the ground; on fine evenings my choice is a Fujitsu Stylistic Tablet PC, which has a more delicate constitution but is somewhat lighter in the hand.

Digital drawing has a number of advantages over traditional sketching. Observing templates can be pre-loaded on the computer ready to be drawn upon, or if necessary prepared at the eyepiece using the appropriate software, avoiding the frustrating business of having to prepare physical outline blanks to be sketched upon. One useful method for those with WiFi-enabled PDAs is to access the accurate and versatile NASA JPL Solar System Simulator on the Internet (I use a freeware astronomy program called Taiyoukei to do this). The simulator enables a real-time, suitably scaled and phase-depicted image of Mercury or Venus to be generated, which in turn can be screen captured and transferred to a drawing program, to be used as an observational drawing template.

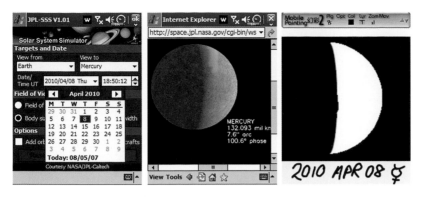

Using the NASA JPL Solar System Simulator on PDA to generate an observing blank of Mercury. This shows the planet at its favourable greatest elongation east on 8 April 2010. Peter Grego.

PDA, UMPC and tablet PC screen are backlit and brightness adjustable, so there's no need to hold a torch in one hand while making the sketch. There are many excellent graphics programs available which have great versatility in terms of the media that they replicate (from pencil effects to oil paints) the range of tones able to be reproduced and a broad spectrum of effects that can be applied to parts of the drawing, such as blending, smoothing, blurring, tone adjustment and so on. Unlike pencil sketching, errors can be easily remedied by simply going back a step. As a keen astronomical cybersketcher myself, I find that publication-ready drawings can be produced at the eyepiece on PDA or tablet PC without having to prepare a 'neat' sketch afterwards, or at least with the minimum of retouching at the computer indoors.

Written Notes

All observational drawings should be accompanied by notes, made on the same sheet of paper, stating the name of the planet (observations of Mercury and Venus can sometimes look confusingly similar), along with the date, observation start and finish times (in Universal Time), the instrument and magnification employed, integrated light or filters used, and the seeing conditions. Short written notes, made at the telescope eyepiece, may also point out any unusual or interesting features that have been observed or suspected, but which may not necessarily be obvious or able to be depicted on your drawing. ALPO's Venus observing sheet contains a very handy tick-off checklist of observational phenomena, such as the degree of terminator shading, atmospheric features, and so on, which, in addition to reducing the amount of time spent writing notes, prompts the observer to systematically look out for specific features while observing.

Date and Time: Amateur astronomers around the world use UT (Universal Time), which is the same as GMT (Greenwich Mean Time). Observers need be aware of the time difference introduced by the world time zone in which they reside and any local daylight savings adjustments to the time, and convert this to UT accordingly – the date should be adjusted too. Times are usually given in terms of a 24-hour clock – for example, 3.25 p.m. UT can be written as either 15:25, 1525 or 15 h 25 m UT.

Seeing: To estimate the quality of astronomical seeing, astronomers refer to one of two scales of seeing. In the UK, many observers use the **Antoniadi Scale**, devised specifically for lunar and planetary observers:

AI – Perfect seeing, without a quiver. Maximum magnification can be used if desired.

AII – Good seeing. Slight undulations, with moments of calm lasting several seconds.

AIII – Moderate seeing, with large atmospheric tremors.

AIV – Poor seeing, with constant troublesome undulations.

AV – Very bad seeing. Image extremely unstable, hardly worth attempting to observe, since even the planet's phase may not be able to be discerned.

In the United States, seeing is often measured from 1 to 10 on the **Pickering Scale**. The scale was devised according to the appearance of a highly magnified star and its surrounding Airy pattern through a small refractor. The Airy pattern, an artefact introduced by optics, will distort according to the degree of atmospheric turbulence along its light path. Under perfect seeing conditions, stars look like a tiny bright point surrounded by a complete set of perfect rings in constant view. Of course, most planetary observers don't check the Airy pattern of stars each time they estimate the quality of seeing during an observing session – an estimate is made, based on the steadiness of a bright stellar image.

P1 – Terrible seeing. Star image is usually about twice the diameter of the third diffraction ring (if the ring could be seen).

P2 – Extremely poor seeing. Image occasionally twice the diameter of the third ring.

P3 – Very poor seeing. Image about the same diameter as the third ring and brighter at the centre.

P4 – Poor seeing. The central disk often visible; arcs of diffraction rings sometimes seen.

P5 – Moderate seeing. Disk always visible; arcs frequently seen.

P6 – Moderate to good seeing. Disk always visible; short arcs constantly seen.

P7 – Good seeing. Disk sometimes sharply defined; rings seen as long arcs or complete circles.

P8 – Very good seeing. Disk always sharply defined; rings as long arcs or complete but in motion.

P9 – Excellent seeing. Inner ring stationary. Outer rings momentarily stationary.

P10 – Perfect seeing. Complete diffraction pattern is stationary.

Considerable confusion can be caused if a simple figure is used to estimate seeing, without indicating whether it's made on the Antoniadi or Pickering scale. So, in addition to designating the seeing with a letter and a figure (AI to V or P1 to 10), a brief written description of seeing, such as 'AII – Good with occasional moments of excellent seeing' can be made.

Conditions: An indication of the prevailing weather conditions, such as the amount of cloud cover, the degree and direction of wind and the temperature.

Transparency: The quality of atmospheric clarity, known as transparency, varies with the amount of smoke and dust particles in the atmosphere, along with cloud and haze. Industrial and domestic pollution causes transparency to be worse in and around cities. A transparency scale of 1 to 6 is often used, according to the magnitude of the faintest star detectable with the unaided eye.

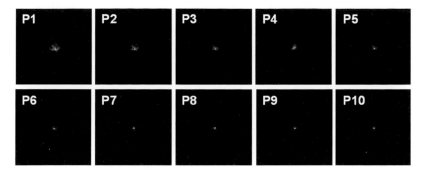

The Pickering Scale of seeing is based on the quality of a bright magnified star's image. Peter Grego.

Mercury and Venus Data

Much of the necessary information about Mercury and Venus is published in annual astronomical ephemerides such as the *Handbook of the British Astronomical Association*, the *Astronomical Almanac* (a joint publication of the US Nautical Almanac Office and Her Majesty's Nautical Almanac Office in the UK) and the Multiyear Interactive Computer Almanac (MICA, published by the US Naval Observatory). Typical tabulated data for Mercury and Venus given in astronomical ephemeredes includes:

Date(s) of greatest elongation west (morning apparition) and extent of elongation (in degrees W); date(s) of superior conjunction.

Date(s) of greatest elongation east (evening apparition) and extent of elongation (in degrees E); date(s) of inferior conjunction.

Date (given in equally-spaced increments, e.g. 5 day intervals).

Planet's celestial co-ordinates in RA (Right Ascension) and Dec (Declination).

Magnitude (to nearest 1/10 of magnitude).

Diameter (in arcseconds).

Phase (given as a percentage or as a decimal number, e.g., 50% or 0.5).

Elongation (in degrees, E expressed as a positive and W as a negative value).

Central Meridian (CM, often not included in Venus data because the planet's surface is not visible in integrated light).

Distance from the Earth in Astronomical Units (AU).

Computer Programs

Many astronomical programs for personal computers and PDAs are capable of giving the user much more than a few dry figures. I haven't tried them all, but I can recommend both the *Starry Night* and *RedShift* software suites for PC; different versions of these programs have been released in recent years, but they all have numerous excellent features for the planetary observer and the amateur astronomer in general.

PDA (Personal Digital Assistant) astronomy programs are very useful to access at the eyepiece. My favourite astronomical PocketPC programs include *TheSky*, *Pocket Universe* (both commercial programs with downloadable demos) and *Taiyoukei* (a very impressive freeware program with a lovely user interface). Among the numerous Palm PDA astronomy programs, one of the most capable of these is the shareware program *Planetarium* by Andreas Hofer. As mentioned above, the NASA JPL Solar System Simulator is an excellent Internet resource, accessible a number of ways, including via Internet on PDA or PC. It can be found at http://space.jpl.nasa.gov/

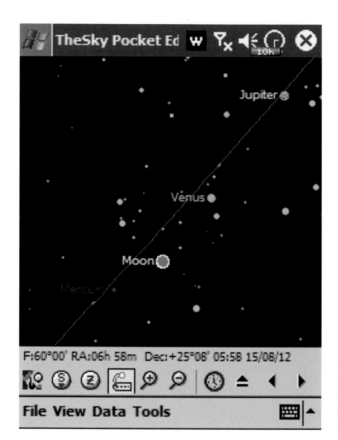

TheSky software for PocketPC. The view shows the planetary grouping on the morning of 15 August 2012. Image attached.

With their sensitive electronics and LCD (liquid crystal display) screens, PDAs have an optimum operational temperature range equivalent to the comfort range experienced by the human body – they work best at room tempeature, and like our bodies, PDAs are less efficient under extremes of hot or cold. Used in the field for night time astronomy, cold is the main factor affecting a PDA's functioning – the first component to suffer is the LCD screen, accompanied by an increasing number of random errors as the temperature falls.

Having extensively used PDAs at the telescope eyepiece for several years, now, including numerous sessions with sub-zero temperatures, I can recall only one instance of PDA failure (and that was a software glitch). If a PDA is kept in a loose inner coat pocket while observing, it won't get the chance to succumb to the cold, as body heat will prevent it from freezing up. When in use, hand warmth will help raise its temperature; you need a bare hand, or naked fingertips at the very least, to

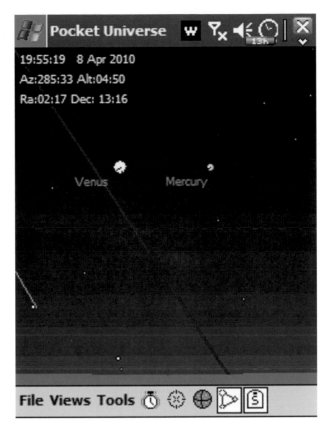

Pocket Universe software for PocketPC. The view shows Mercury and Venus in the evening skies of 8 April 2010.

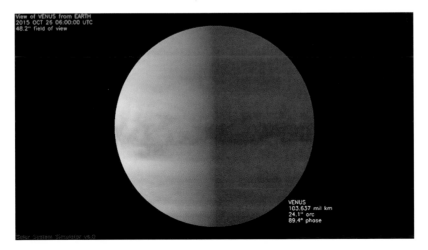

NASA JPL Solar System Simulator, an Internet-based program, allows accurate graphics of Mercury and Venus to be generated. This shows Venus at its favourable greatest elongation west on 26 October 2015.

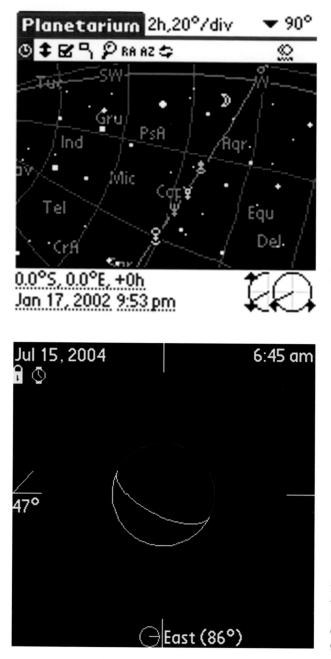

Screenshot of AHo Software's Planetarium, a highly capable astronomy program for Palm OS.

2sky, a versatile planetarium program for Palm OS, is capable of displaying accurate phased graphics of Venus.

use the PDA's stylus, which is always on the small and spindly side. Importantly, don't place your PDA on any exposed surface outside for any length of time, as it is likely to rapidly dew over and/or freeze up. A PDA that has cooled to a temperature of below –5°C will perform poorly, and if allowed to cool to below –15°C it won't work at all.

Line Drawing

Outline drawings may be an alternative or adjunct to making shaded pencil drawings at the eyepiece. Annotated line drawings can be every bit as informative as a shaded pencil sketch, but the technique should not be considered a quick and easy alternative to tonal drawing, since they ought to be drawn just as carefully and with the same amount of attention to detail. Intensity estimates are a valuable addition to line drawings. This requires the observer to estimate of the brightness of each distinct area depicted on the drawing, using a scale of 0 to 5.

Intensity Estimates Scale

0 Extremely bright areas.
1 Bright areas.
2 General hue of the planet's disk.
3 Faint shading near the limit of visibility.
4 Easily discernable shading.
5 Unusually dark shading.

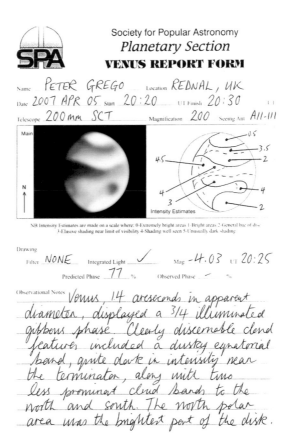

Example of a completed Venus report form (SPA Planetary Section) with a tonal drawing and intensity estimate. Peter Grego.

The eye is capable of differentiating between hundreds of shades of grey, so a skilled observer can easily subdivide the basic scale yet further. Unlike, say variable star estimations, these tend to be qualitative visual estimates, rather than quantitative ones. Additionally, optical illusions are capable of playing tricks with the observer – two areas of true identical tone may appear to have widely different tones when set next to areas of different brightness and contrast.

Copying-up Your Observations

Few people have the ability to produce completely error-free observational drawings at the eyepiece. Therefore it is best to prepare a neat copy of your observational drawing as soon after the observing session as practically possible, while the scene remains vivid in your mind, since accurate recall tends to fade pretty quickly. A fresh drawing prepared indoors will be far more accurate than the original telescopic sketch that it is based upon, as the observer will be able to recall little things about the original sketch that perhaps weren't quite right and needed to be rectified on the neat drawing. Unlike capturing an image on CCD, many of the details of the observation are contained inside the observer's head and are accessible only using the mind's eye.

It is perfectly possible to translate the most rough and ready drawing made at the eyepiece into a far more accurate and pleasing depiction of what was observed. Some of my own at-the-eyepiece observational drawings (some, I'm embarrassed to admit, were made freehand and in ballpoint pen on lined paper owing to a lack of resources while in the field) are pretty worthless in themselves, inaccurate and quite horrible to look at. However, their real value is a transient thing, accessible only to the observer a few hours after the observing session and only capable of being revealed in a fresh, neat observational drawing. The neat copy of your original telescopic sketch can be used as the template for further drawings, or used to electronically scan or photocopy.

Copied drawings may be prepared in a variety of media. Superb results can be achieved using India ink washes, and paintings in gouache – a watercolour medium that it is possible to apply fairly thickly and in a controlled manner – are excellent for reproducing observation on a larger scale for exhibition purposes. Both these techniques require proficiency in brushwork, and although a description of the methods involved is beyond this book, practice, experimentation and perseverance will pay big dividends.

When making physical drawings, soft pencil on smooth white paper is by far the quickest and least fussy medium. Once completed, pencil drawings need to be sprayed with a fixative so that they don't smudge if they are inadvertently rubbed. Regular copy shop photocopies of tonal pencil drawings are not quite good enough to submit to astronomical society observing sections or for publication in magazines, as the full range of tones in the drawing will not be captured, and it may appear somewhat dark and grainy. These days, however, most planetary observing sections are happy to accept a high quality laser print or a digitally scanned drawing. Some commercial magazines may insist on having the original artwork to work from, or at the very least for high quality, high resolution scans to be submitted on floppy disc or by email.

Tempting though it may be to discard old drawings, they represent a permanent record of your observations and hard work at the eyepiece. At the very least, comparing old observations with more recent ones will demonstrate how much your skills of observation and recording have improved. Original drawings can be used as the basis for subsequent copies for the observing sections of any astronomical societies to which you belong, or for publication. For these reasons, do hang on to your original drawings for future reference. Devote a folder or a ring-binder to your original drawings of Mercury and Venus, placing each in clear polythene pockets a clean, dry environment.

Imaging Mercury and Venus

Conventional Photography

A beautiful bright, pin-sharp image of Venus can often be viewed using a small telescope, so it may seem reasonable to suppose that a camera is capable of capturing the same image with little difficulty. Venus is certainly bright enough to register on just about any conventional film camera pointed through a telescope eyepiece, but successful planetary photography using an ordinary film camera – be it a simple compact camera or a 35 mm SLR (single lens reflex) – is a rather more difficult and involved process than many imagine.

An equatorially driven telescope and a firmly fixed camera, keeping the planet steadily centred in the field of view, are essential to capture Venus on conventional film. While visual observers can contend with the movement of a planet as it drifts through the field of view of an undriven telescope, the slightest movement will produce motion blurring in a photographic image; the longer the exposure, the greater the degree of blurring.

Afocal Photography

Afocal photography is the process of photographing an object through the telescope eyepiece with a non-SLR compact film or digital camera. Basic compact film and digital cameras have fixed lenses that are usually preset to focus objects ranging from a few metres away to infinity; with their preset exposures for standard photography, they are unable to be fine-tuned for astronomical purposes. They are however suitable for imaging a bright planet like Venus.

Non-SLR film cameras have a viewfinder that is slightly to one side of the photographic optical axis – fine for composing quick and easy images of everyday scenes, but completely unusable for afocal photography, as it will not show the image of Venus being projected into the camera. Given that the view through the camera will not be visible during afocal photography with regular film cameras, Venus must be lined up in the crosshairs of an accurately aligned finderscope and photographed without the imager being able to see the picture's composition.

Afocal photography through most compact digital cameras is somewhat easier because the image being projected onto the camera's CCD chip appears on an electronic display – albeit a somewhat coarse and small LCD screen – at the rear of the camera. Digital cameras are designed for everyday use, so their automatic settings may pose considerable problems when attempting to photograph Venus, so experimentation with the camera's various settings is necessary to produce the best results.

First, Venus is focused by eye through the telescope eyepiece, and the camera is then positioned close to the eyepiece and held there firmly. If there is a focus

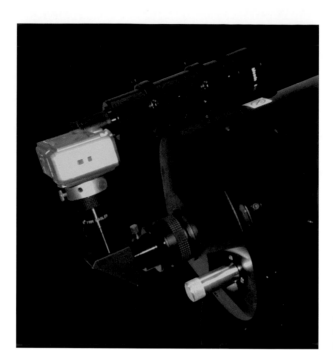

A digital camera set up for afocal imaging with a 200 mm SCT. Peter Grego.

adjustment on the camera, it's best set to infinity. Some basic cameras are meant to be used without any accessories, so some kind of makeshift camera adapter will need to be constructed to mate it to the telescope – many cameras are lightweight enough to be temporarily fixed to a telescope with a little blu-tack and electrical tape. If your camera has a standard tripod bush at the bottom of the body, it will be possible to mount the camera more securely or attach it to a commercially available telescope camera mount.

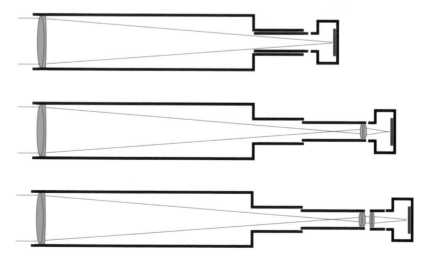

Three imaging techniques – prime focus, eyepiece projection and afocal imaging. Peter Grego.

Conventional SLR Photography

SLR cameras direct the light from the object being photographed through the main camera lens to the eye via a mirror, prism and eyepiece. The area framed in the viewfinder is exactly the area that will be included in the final image. When the shutter button is pressed, the mirror instantly flips out of the way of the light path, allowing it to project directly onto the film.

SLR cameras have grown increasingly sophisticated since their general introduction in the 1960s. Top-of-the-range SLRs are provided with electronic controls that can automatically adjust every aspect of the camera's functions. A cable release will minimise the effects of vibration that occurs when the shutter button is pressed; better still, most SLRs can be adjusted to take an exposure after a timed delay, giving the vibrations in the telescope time to damp down after the camera is handled. In some basic SLR models, the movement of the shutter can vibrate the camera and telescope, causing a little blurring of the image – this would of course present more of a problem when taking high magnification images, as the effects of any movement of the telescope would be exacerbated.

The author attempts SLR photography of Venus. Peter Grego.

Some of the most stunning images of Mercury and Venus have been wide angle twilight views taken with SLR cameras piggybacked on a telescope or even mounted on an undriven tripod. The composition of such images is very much a matter of taste, but there are a few basic rules to taking a wide-angle image of sufficient interest and spectacle to merit its inclusion on the *Astronomy Picture of the Day* website (http://antwrp.gsfc.nasa.gov/apod/astropix.html). First, include an interesting foreground – say, a historical site, an area of natural beauty or a panoramic cityscape. Reflections of bright celestial objects like Venus set off a picture nicely, so you might like to search for a suitable smooth-surfaced pond, lake or sea. Events such as lunar, stellar and planetary appulses make an image really memorable, so keep a look out for suitable photographic opportunities by scanning the sky diary pages of your favourite astronomy magazine or by running your PC astronomy program.

Photograph of the very close encounter between Mercury and Venus on the evening of 27 June 2005, when the two planets were just 5 arcminutes apart. The phases of both planets are evident. Chris Dignan.

Selected Large Scale Photo Opportunities Featuring Mercury and Venus, 2009–2019

Please note that the following list is just a sampling of opportunities to view and/or photographically capture some visually pleasing inner planet action – it's by no means a comprehensive, all-encompassing list of the best photographic opportunities in the coming decade. Transits of both Mercury and Venus have been left out of this listing (see the chapters on Mercury and Venus for

2009 January 22. Evening. Venus (mag –4.5) near GEE and Uranus (mag 5.9) just 1° apart).

2009 April 26. Evening (N hem favoured). Mercury (mag 0.1) at GEE (20°), featuring a lovely appulse with the young crescent Moon (2° to the north) and the Pleiades.

2009 May 1. Evening (N hem favoured). Mercury 1° east of the Pleiades.

2009 May 21. Morning (S hem favoured). Venus (mag –4.4) in close grouping with Mars (mag 1.1) and the waning crescent Moon.

2009 October 8. Morning (N hem favoured). Mercury (mag –0.7) just 19 arcminutes south of Saturn (mag 1.1), with Venus (mag –4.9) some 6° to the west.

2010 August 13. Evening (S hem favoured). Beautiful gathering of Mercury (mag 0.6), Venus (mag –4.3), the young crescent Moon, Mars (mag 1.5) and Saturn (mag 1.1). Excellent from far southern hemisphere locations, South Australia and New Zealand, unfavourable from northern temperate latitudes.

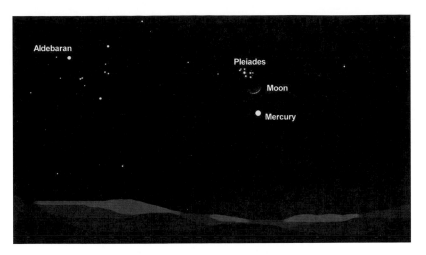

2009 April 26 (view from UK).

2011 March 27. Morning (S hem favoured). Venus (mag −4.0) just 20 arcminutes away from Neptune (mag 7.9).

2011 May 7. Morning (S hem favoured). Close gathering of Mercury (mag 0.6), Venus (mag −3.9), Mars (mag 1.3) and Jupiter (mag −2.1). Excellent from far southern hemisphere locations, South Australia and New Zealand, unfavourable from northern temperate latitudes.

2012 February 9. Evening (N & S hem). Venus (mag −4.1) within half a degree of Uranus (mag 5.9), with Jupiter (mag −2.3) further east.

2012 March 5. Evening (N hem favoured). Mercury (mag −0.3) at GEE (18°), with Venus (mag −4.2) and Jupiter (mag −2.2) further to the east.

2012 April 3. Evening (N hem favoured). Venus (mag −4.4) a spectacular sight amid the Pleiades.

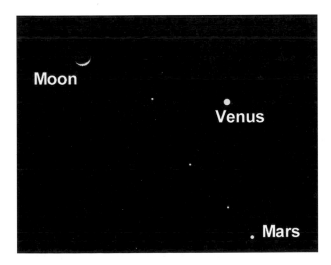

2009 May 21 (view from South Australia).

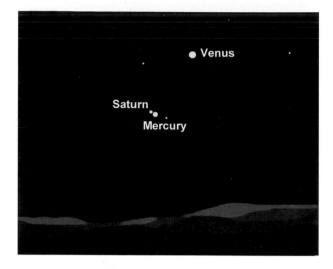

2009 October 8 (view from UK).

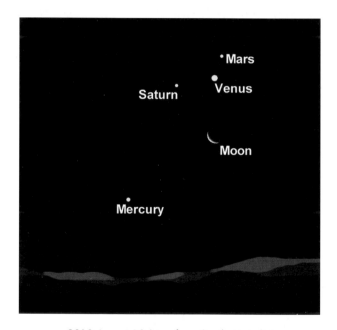

2010 August 13 (view from South Australia).

2012 July 15. Morning (N & S hem). Venus (mag −4.5) grouped with Jupiter (mag −2.1) and the waning crescent Moon amid the Hyades star cluster.

2012 December 2. Morning (N & S hem). Mercury (mag −0.3) near GEW, in line with Venus (mag −4.0) and Saturn (mag 0.7) further west. Followed on 11 December by a close approach of the waning crescent Moon.

2013 October 7. Evening (S hem favoured). Mercury (mag 0.0) near GEE, near Saturn (mag 0.6), young crescent Moon and Venus (mag −4.2) further east.

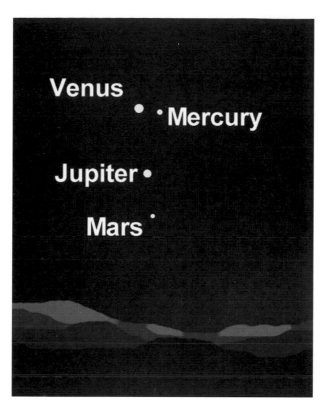

2011 May 7 (view from South Australia).

2013 November 25. Morning (N hem). Mercury (mag −0.7) near Saturn (mag 0.6) and Comet Encke (mag 4.5), challenging.

2014 August 19. Morning (N hem favoured). Venus (mag −3.9) within one degree of Jupiter (mag −1.8), set against the background of the star cluster Praesepe.

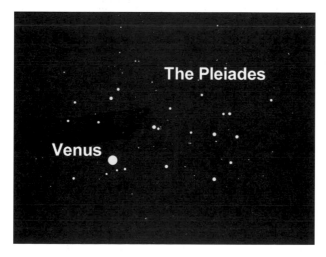

2012 April 3 (view from the UK).

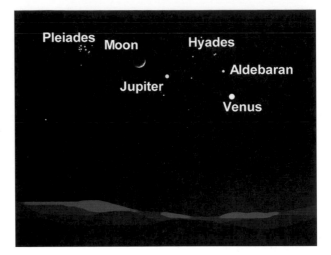

2012 July 15 (view from South Australia).

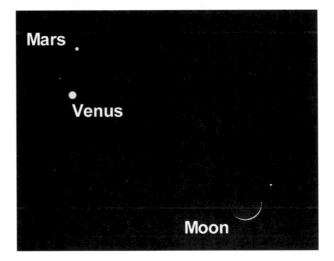

2015 February 20 (view from the UK).

2015 January 14. Evening (N & S hem). Mercury (mag −0.6) at GEE just 1 away from Venus (mag −3.9).

2015 February 20. Evening (N hem favoured). Venus (mag −4.0) within a degree of Mars (mag 1.3) and near the waxing crescent Moon.

2015 October 16. Morning (N hem favoured). Mercury (mag −0.6) at GEW, with a close grouping of Venus (mag −4.4), Mars (mag 1.8) and Jupiter (mag −1.8) further west (the trio being closest on October 24).

2016 January 7. Morning (N & S hem). Close grouping of Venus (mag −4.0), Saturn (mag 0.5) and the waning crescent Moon

2016 February 6. Morning (N hem favoured). Mercury (mag −0.1) near GEW, close to Venus (mag −4.0) and the waning crescent Moon.

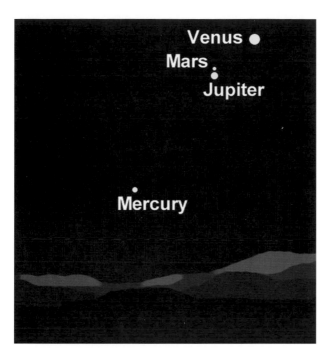

2015 October 16 (view from the UK).

2017 January 12. Evening (N & S hem). Venus (mag −4.4) at GEE is within half a degree of Neptune (mag 7.9), while Mars (mag 1.0) lies further east.

2017 September 12. Morning (N hem favoured). Mercury (mag −0.2) near GEW, close to Mars (mag 1.8), with Venus (mag −3.9) further west. On September 17

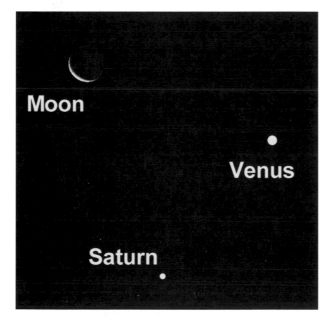

2016 January 7 (view from the UK).

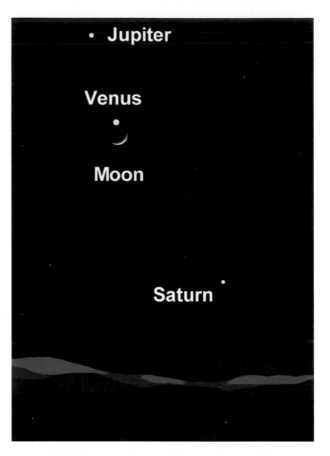

2019 January 31 (view from South Australia).

Mars and Mercury are just 19 arcminutes apart, with the waning crescent Moon approaching Venus.

2017 November 23. Evening (S hem favoured). Mercury (mag −0.3) at GEE, 5° from Saturn (mag 0.5), with the young crescent Moon further to the east.

2018 March 15. Evening (N hem favoured). Mercury (mag −0.3) and Venus (mag −3.9) just 4° apart. The slender waxing crescent Moon glides by on March 19.

2018 December 22. Morning (N & S hem). Mercury (mag −0.4) and Jupiter (mag −1.8) less than 1° apart. Venus (mag −4.6) lies further west.

2019 January 31. Morning (S hem favoured). Venus (mag −4.3) in line with the nearby waning crescent Moon, Jupiter (mag −1.9) to the west and Saturn (mag 0.6) to the east.

2019 April 12. Morning (N hem favoured). Mercury (mag 0.4) at GEW and Venus (mag −3.9) are less than 5° apart.

2019 June 18. Evening (unfavourable from far N hemisphere) Mercury (mag 0.2) and Mars (1.8) approach within 14 arcminutes of each other.

2019 October 29. Evening (S hem favoured). Grouping of Mercury (mag 0.3), Venus (mag −3.9) and the young crescent Moon, with Jupiter (mag −1.9) located further to the east.

Mercury, imaged by Jamie Cooper on 29 November 2006.

GEE – Greatest elongation east
GEW – Greatest elongation west
N hem – Northern hemisphere
S hem – Southern hemisphere

Venus (upper star) and Jupiter make a very close approach in the South African skies of 3 September 2005. Image by Jamie Cooper.

Film Types

A film's ISO rating indicates its 'speed' – the higher the ISO, the faster the film and less exposure time is required. 200 ISO is a medium speed film, and inexpensive generic 200 ISO colour print film is fine for beginning planetary photographers to experiment with. Quality tends to vary from brand to brand, and even varies among different batches of the same 'budget' brand. Slower films have finer grains allowing more detail to be captured, while grain size increases with a film's ISO rating. This may not be evident in comparing regular sized photographic prints, but enlargements will clearly show the difference. Photographs taken with slower film can withstand much more enlargement than those taken with faster film, but slow films also have the drawback of requiring more time to expose. High magnification shots of the planets made on regular film require an accurately driven equatorial drive. Specialist Kodak Panatomic X (medium-to-high contrast, extremely fine grain and resolving power), Kodak Plus-X (medium-speed ISO 125, extremely fine grain and excellent sharpness) and Kodak Tri-X (fine grain, high sharpness and resolving power) are all films sensitive in the ultraviolet wavelengths, enabling Venusian cloud detail to be captured.

Prime Focus Photography

When a camera body (a camera minus its lens) is attached to a telescope (minus its eyepiece), the light falling onto the film is at the main focal point of the telescope's objective lens (or mirror). Conventional prime focus photography, even when made with a long focal length telescope, delivers a terribly small planetary image on a 35 mm film frame. A Barlow lens (standard ones come in ×2 and ×3 varieties) will effectively increase the focal length of a telescope, producing an enlarged image. Focusing a planet at prime focus is done by looking through the camera's viewfinder, using the telescope's own focuser. The process is usually quite forgiving, and by no means as exacting as focusing an image using eyepiece projection (see below).

Prime focus planetary imaging is extensively used with webcams and astronomical CCD cameras. CCD chips are tiny in comparison to 35 mm film, so the effective magnification they deliver is far larger (see below for CCD photography). Precise focusing is critical, and best achieved using an electronic focuser rather than manual knob-twiddling.

Eyepiece Projection

High magnification planetary photographs can be achieved by inserting an eyepiece into the telescope and then projecting the image into the camera, minus its lens. Adapters are widely available that fit into standard sized eyepiece holders (1.25-inch and 2-inch diameters) and the bodies of various makes of SLR. Orthoscopic and Plössl eyepieces deliver crisp images with flat fields of view. Eyepiece projection will deliver a far higher magnification image than prime focus photography, the degree of magnification depending on the focal length of the telescope and the eyepiece, and the distance of the eyepiece from the film plane. Shorter focal length eyepieces will deliver higher magnifications, and increasing the distance of the eyepiece from the film plane will also increase magnification. Focusing is achieved by looking directly at the magnified planetary image through the camera viewfinder and adjusting the telescope's focusing knob until the planet appears sharply focused.

Digital Imaging

A CCD (charge coupled device) is a small flat chip – about the diameter of a match head in most commercial digicams – made up of an array of tiny light sensitive elements called pixels. Light hitting each pixel is converted to an electrical signal, and the intensity of this signal directly corresponds to the brightness of the light striking it. This information can be stored digitally in the camera's own memory or transferred to a PC, where it can be processed into an image. CCDs in the lowest-end digital cameras may have an array of 640 × 480 pixels; an avergae 4 megapixel camera will have a 2240 × 1680 pixel CCD, while a higher-end camera with a 3072 × 2304 pixel chip boasts 7.2 megapixels.

Webcams and dedicated astronomical CCD cameras have enabled amateur astronomers with quite modest equipment the opportunity to obtain very satisfying images of Mercury and Venus. Digital images are infinitely easier to enhance and manipulate than a conventional photograph in a photo lab darkroom. Although just about anyone can take an acceptable planetary snapshot by pointing a CCD camera through a small telescope, it requires considerable skill and expertise – both in the field, and later at the computer – to produce high resolution images that show surface detail.

Camcorders

Camcorders have fixed lenses, and footage must be obtained afocally through the telescope eyepiece. The same problems that affect afocal imaging using conventional film and digital cameras apply to camcorders. Camcorders tend to be heavier than digital cameras, and it is essential that the camcorder is coupled to the telescope eyepiece as sturdily as possible. Some of the same equipment designed to hold a digital camera in place when taking afocal planetary images can be used to secure a lightweight camcorder to a telescope.

Digital camcorders are the lightest and most versatile camcorders, and their images can be easily transferred to a computer for digital editing using the same techniques as images obtained with a webcam (see below). Once downloaded onto a computer, individual frames from digital video footage can be sampled individually (at low resolution), stacked using special software to produce detailed, high resolution images, or assembled into clips that can be transferred to a CD-ROM, DVD or videotape. The process can be time consuming – the time spent running through the video footage and processing the images may amount to far longer than the time that was actually spent taking the footage. Digital video editing also consumes a great deal of a computer's resources, in both terms of memory and storage space – the faster a computer's CPUs and graphics card, the better. At least 5 gigabytes need to be available on your computer's hard drive for the most basic editing of video clips.

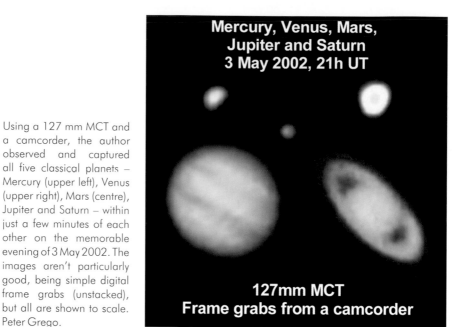

Mercury, Venus, Mars, Jupiter and Saturn 3 May 2002, 21h UT

127mm MCT
Frame grabs from a camcorder

Using a 127 mm MCT and a camcorder, the author observed and captured all five classical planets – Mercury (upper left), Venus (upper right), Mars (centre), Jupiter and Saturn – within just a few minutes of each other on the memorable evening of 3 May 2002. The images aren't particularly good, being simple digital frame grabs (unstacked), but all are shown to scale. Peter Grego.

Webcams

Although they are usually designed for use in the home to enable communication between individuals over the Internet, webcams can be used to capture high resolution images of the planets. Lightweight and versatile, webcams are just a fraction of the cost of dedicated astronomical CCD cameras. Just about any commercial webcam hooked up to a computer and a telescope can be used to image Mercury and Venus, although the quality of the images depends on a number of skills and techniques that may only be improved through patience, practice and perseverance.

While webcams may not have as sensitive a CCD as more expensive astronomical CCD cameras, their ability to record video clips made up of hundreds, or thousands of individual images gives them a distinct advantage over the single-shot astronomical CCD. By taking a video sequence made up of dozens, hundreds or even thousands of individual frames, the effects of poor seeing can be overcome by processing only the clearest images in the video clip. These images – selected either manually or automatically – can then be combined using stacking software to produce a highly detailed image. This may show as much detail as that seen visually through the eyepiece using the same instrument.

Webcams are usually used at the telescope's prime focus (minus the webcam's original lens) to image Mercury and Venus. A number of webcams, such as the popular Philips ToUcam Pro, have easily removable lenses; commercially available telescope adapters can be screwed in their place, permitting easy attachment to a telescope. Some webcams however require disassembly to remove the lens, and the adapter needs to be home made.

CCD chips are sensitive to infrared light, and the original lens assembly usually contains an infrared blocking filter – without the filter, a really clean focus is not possible, since infrared is focused differently to visible light. IR blocking filters are however available to fit into the telescope adapter, allowing only visible light wavelengths to pass through to a sharp focus.

Like the CCD chips in most other digital devices, webcam CCDs are very small – 3.6 × 2.7 mm in a Philips ToUcam Pro. Used at the telescope's prime focus, the magnification produced on a webcam is usually augmented with a Barlow lens. Because of the high magnifications produced by a webcam at prime focus on an average amateur telescope, a well polar-aligned driven equatorial telescope with electric slow motion controls is essential in order to capture a relatively static video clip lasting ten or twenty seconds. If the image drifts too much during the clip, the software used to process the video clip may not be able to produce a good alignment.

Focusing a webcam accurately can prove time consuming, but to achieve a rough focus, it is best to set up during the daytime and focus on a distant terrestrial object using the telescope and webcam, viewing the laptop monitor and adjusting the focus manually. Once the terrestrial object has been focused, lock the focus or mark the focusing barrel with a CD marker pen.

During the imaging session, the planet is centred in the field of view using the telescope's finderscope, and if it is accurately aligned the planet will appear on the computer screen, probably in need of further focusing. When the telescope's focus is adjusted manually, care must be taken not to nudge the instrument too hard, as the planet under scrutiny may disappear altogether out of the small field of view. Patient trial and error will eventually produce a reasonably sharp focus – once achieved, lock the focuser and mark the focusing tube's position so that an approximately sharp focus can be found quickly during subsequent imaging sessions.

Achieving a good focus makes the difference between a good planetary image and a great one – a fraction of a millimetre can make the difference between a good focus and a tack-sharp one. Manual focusing in the manner outlined above is exceedingly time consuming, and a perfect focus is more likely to be found by chance than trial and error. Electric focusers enable the focus to be adjusted remotely from the telescope, and deserve to be considered an essential accessory to the planetary imager. Electric focusers save much time and make a great difference to your enjoyment of imaging; importantly, they offer infinitely more control over fine focusing. A webcam attached to a computer through a high speed USB port will deliver a rapid refresh rate of the image, enabling fine focusing in real time.

Video sequences of the planets can be captured using the software supplied with the webcam. It is necessary to override most of the software's automatic controls – contrast, gain and exposure controls require adjusting to deliver an acceptable image. Many imagers prefer to use black and white recording mode, which cuts down on signal noise, takes up less hard drive space and eliminates any false colour that may be produced electronically or optically.

The webcam's greatest strength is the sheer number of images that it provides in a single video capture. Single-shot dedicated astronomical CCD cameras costing ten times as much as a webcam are capable of taking just one image at a time; while this image may have far less signal noise and a higher number of pixels than one taken with a webcam, in mediocre seeing conditions the chances that the image was taken at the precise moment of very good seeing are small. Webcams can be used even in poor seeing conditions, as a number of clearly resolved frames will be available to use in an extended video sequence. Video sequences are usually captured as AVI (Audio Video Interleave) files.

Most webcams are able to record image sequences with frame rates of between 5 and 60 fps (frames per second). A ten second video clip made at 5 fps will be composed of 50 individual frames and may take up around 35 Mb of computer memory. At 60 fps there will be 600 exposures, and the amount of space taken up on the hard drive will be proportionately greater. Using the webcam's highest resolution (in most cases, an image size of 640 × 480) a frame rate of between 5 and 10 frames per second and a video clip of 5 to 10 seconds is optimum. If five to ten of these clips are secured in quick succession, the imager will have between 125 and 1,000 individual images from which to work from. While higher sample sizes produce better results, the time spent in processing them increases too – a fact which must be considered if making manual selections of suitable frames. Since the actual increase in quality is found to rise on a logarithmic scale, there comes a point when no visible improvements in a planetary image can be seen, regardless of the size of the sample.

Astronomical image editing software is used to analyse the video sequence, and there are a number of very good freeware imaging programs available. Some programs are able to work directly from the AVI, and much of the process can be set up to be automatic – the software itself selects which frames are

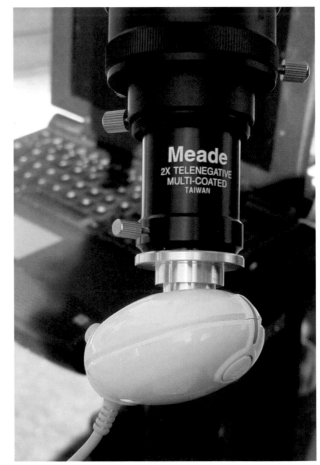

A Philips PCVC740 webcam set up with a Barlow lens at prime focus for lunar and planetary imaging. Peter Grego.

the sharpest, and these are then automatically aligned, stacked and sharpened to produce the final image. If more control is required, it is possible to individually select which images out of the sequence ought to be used – since this may require up to a thousand images to be visually examined, one after another, this can be a laborious process, but it often produces sharper images than those derived automatically.

Images can be further processed in image manipulation software to remove unwanted artefacts, to sharpen the image, enhance its tonal range and contrast and to bring out detail. Unsharp masking is one of the most widely used tools in astronomical imaging – almost magically, a blurred image can be brought into a sharper focus. Too much image processing and unsharp masking may produce spurious artefacts in the image's texture, shadowing, ghosting and chiaroscuro effects and a progressive loss of tonal detail. Each individual imager tends to develop their own particular methods of enhancing their raw planetary images. Incredible though it may seem, it is quite possible for the top few dozen of the world's most experienced amateur planetary imagers to tell each others' work apart because of the subtle differences displayed in the final image that have been introduced by different combinations of processing techniques.

Ultraviolet and Infrared Imaging of Venus

Some of the most exciting developments in amateur CCD astronomy have involved the increasing and innovative use of various filters to enhance planetary detail that is normally difficult to see visually or beyond the human visual range.

Ultraviolet (UV) imaging is capable of revealing the subtle cloud features of Venus in considerable detail. Since UV wavelengths are absorbed by many types of glass, the best results are obtained by means of prime focus imaging; if a

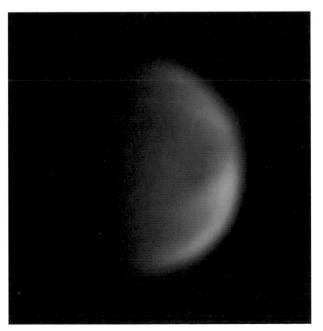

A good example of an amateur UV filtered CCD image of Venus, showing cloud features on a gibbous disk. Imaged on 30 April 2007 by Dave Tyler using a 350 mm SCT and Baader UV filter. Dave Tyler.

diagonal is used it needs to be a mirror type rather than a glass prism. Filters with a maximum transmission of between 3200 Å and 3900 Å deliver the best UV results. Commercially available filters offering this specification include the Kodak W18A filter and the Baader Planetarium U-Filter, the latter having the best overall transmission profile.

The Baader Planetarium U-Filter enhances subtle Venusian cloud detail.

Infrared (IR) imaging of Venus – once the exclusive domain of the well-equipped professional observatory – is a line of investigation being explored by an increasing number of amateur CCD imagers. Venus is an intensely hot planet whose surface and lower atmosphere are hot enough to melt lead. A considerable amount of thermal radiation escapes into space. Cameras tuned to near-infrared wavelengths are capable of recording this red hot glow, allowing the temperature variations of the planet's surface and near-surface environment to be imaged.

IR imaging is however restricted to the night-side of Venus because the cloud tops of the illuminated portion of the planet overwhelm the CCD chip with their sheer brightness. It follows that the best images of the IR glow of Venus' unilluminated hemisphere are to be obtained when the planet is a narrow crescent phase in the few months prior to and following inferior conjunction with the Sun – it happens that Venus attains its largest apparent diameter during this period. While the bright illuminated crescent is overexposed in an image lasting several seconds, subtle detail may be glimpsed within the dark side – detail that appears to correspond with the planet's topography. Brighter areas correspond with hotter regions of low elevation, while darker areas correspond with Venus' cooler highland plateaux and mountain ranges.

Chapter 6

Observing Mercury

Phenomena Common to Mercury and Venus

During the course of the year the Sun appears to trace a path against the background constellations – this path is known as the ecliptic. Since all the major planets have orbital planes roughly coinciding with that of the Earth, they all appear to follow paths within a few degrees of the ecliptic. Even the Moon's orbital plane lies close to the ecliptic, inclined to it by some 5°, so that from time to time there are solar and lunar eclipses (hence the name 'ecliptic'), close lunar-planetary approaches (called 'appulses') and lunar occultations of the planets. The planets also regularly appear to make close approaches to the Sun, but the Sun's glare renders such events unobservable to all but the most experienced amateur astronomers with accurate equipment, apart from on those occasions when Mercury and Venus transit the Sun (see below).

Phases

Mercury and Venus are known as the 'inferior' planets because their orbits both lie within that of the Earth, and neither planet appears to stray very far from the Sun. Both Mercury and Venus undergo a sequence of phases during each elongation from the Sun, in addition to changes in their apparent diameter, which can be followed through a small telescope.

Mercury or Venus are at superior conjunction when they lie directly on the far side of their orbit (as seen from the Earth), hidden from view by the Sun's glare. Following superior conjunction the planet edges east of the Sun and soon becomes visible in the evening sky. At this time, a telescope will reveal the planet as a small gibbous disc. As the planet moves further east of the Sun, its phase diminishes until it reaches half phase (called dichotomy), at its greatest elongation from the Sun. The planet then begins to move westwards, towards the Sun, its phase gradually diminishing to a large crescent, until it is lost in the twilight glare.

Inferior conjunction is reached when Mercury or Venus passes between the Earth and the Sun. On most occasions the planet will appear to pass some distance to the north or south of the Sun at inferior conjunction. On rare occasions, however, an inferior planet happens to pass exactly between the Earth and Sun, so that it appears as a tiny circular silhouette that takes several hours to transit across the solar disc.

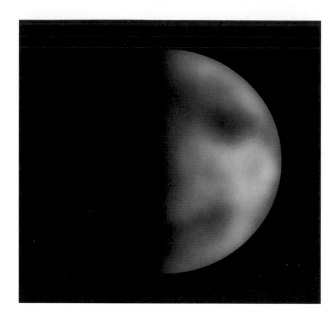

Mercury near dichotomy, from an observation on 8 March 1992. Peter Grego.

Following inferior conjunction the planet moves to the west of the Sun and eventually becomes visible in the pre-dawn skies. Telescopes reveal the post-inferior conjunction planet as a large crescent phase. As time goes by, the planet's phase gradually broadens and its apparent diameter slowly decreases. Dichotomy is reached when the planet is at its greatest elongation west of the Sun. As the planet

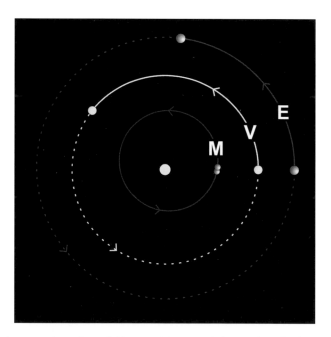

Comparison between the orbits of Mercury, Venus and the Earth, solid lines represent the distance traveled by each planet in one Mercury orbit (88 days). Peter Grego.

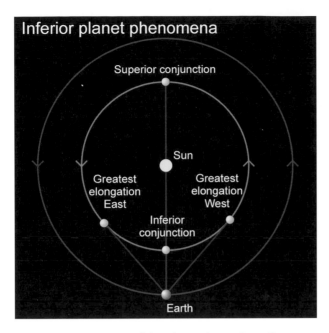

Orbital phenomena of the inferior planets. Peter Grego.

draws closer to the Sun, moving towards the east once more, it becomes a gibbous phase, growing ever smaller in apparent diameter, until it is lost in the Sun's glare approaching superior conjunction.

Lunar Occultations of Mercury and Venus

Since the Moon's orbit is inclined by about 5° to the ecliptic, the area of sky prone to being hidden by the lunar disk lies exclusively in a band of 5° either side of the ecliptic. Both Mercury and Venus, along with all the other major planets, have paths restricted to this near-ecliptic band, and it follows that from time to time the Moon will appear to move directly in front of them, occulting them for a while. Interestingly, Mercury and Venus also occasionally occult stars and planets; bright stellar occultations are rare events, mutual planetary occultations even rarer.

The Moon has no appreciable atmosphere, and the stars are so distant that they appear as point sources of light which disappear and reappear at the edge of the Moon nearly instantaneously when observed through a telescope. Unlike stars, planets have appreciable diameters, and they take a short period of time to be occulted. Mercury's apparent diameter is usually considerably smaller than that of Venus, and occultations are more difficult to observe owing to the planet's proximity to the Sun. Venus is so bright that many daytime occultations can easily be observed through binoculars.

Apparitions

From an observational point of view, Mercury has a reputation as being one of the most elusive planets. Its apparitions are the briefest of all the planets, taking just a couple of months or so to flit between its eastern and western elongations, and its visibility from the Earth is hampered to some extent by its proximity to the Sun.

Because Mercury never appears to stray very far from the Sun, observers in mid-northern locations like the UK, Europe, Canada, the United States, China and Japan are never able to observe the planet set against a truly dark sky background, even at its furthest elongations. Southern hemisphere observers have the advantage of being occasionally able to view Mercury against an astronomically dark sky background during favourable elongations. Additionally, Mercury's rapid angular motion as it curves between superior and inferior conjunction means that it is usually only visible with the unaided eye for a few weeks during each elongation in the most favourable circumstances.

Mercury is best observed when it is at its greatest elongation from the Sun and placed as high as possible above the horizon at sunset or sunrise. From northern temperate regions such as the UK, Europe and much of North America, these circumstances occur during springtime eastern elongations, when the planet is visible in the evening twilight, or during autumnal western elongations when it is visible in the pre-dawn skies. Under such circumstances the imaginary line connecting the Sun and Mercury makes its greatest angle with the horizon at sunset and sunrise respectively. Observers in the southern hemisphere also attain their best

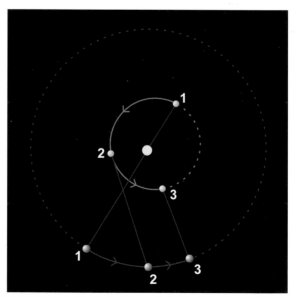

Paths of Mercury and the Earth during early 2016, a typical favourable eastern elongation.
1-1 : Mercury at superior conjunction, 24 March 2016
2-2 : Mercury at greatest elongation East (19°), 18 April 2016
3-3 : Mercury at inferior conjunction (and transit across the Sun) 9 May 2016.

The paths of the Earth and Mercury during early 2016, showing a typical favourable eastern elongation. Peter Grego.

views of Mercury during their own springtime eastern elongations and autumnal western elongations – it follows that an eastern elongation in March will favour the northern hemisphere observer, while a March western elongation in favours the observer in the southern hemisphere.

Mercury's marked orbital eccentricity produces greatest elongations from the Sun which vary between a maximum of 28° (when it is near aphelion) and a minimum of 18° (when near perihelion). Unfortunately for observers located in northern temperate regions or beyond, the best of the greatest elongations coincide with the planet's poorest showing above the sunrise or sunset horizon when the imaginary line connecting the Sun and Mercury makes its shallowest angle with the horizon. For southern hemisphere observers during those same elongations, the angle between Mercury and the Sun at sunrise or sunset is at its greatest, enabling Mercury to be observed higher in the sky for a greater period of time against a darker sky background.

Naked Eye Views

Mercury is at all not difficult to see with the unaided eye if it is observed within a week or so of the date of the most favourable elongations and provided the observer has a reasonably clear sunrise or sunset horizon and a fairly transparent sky. Once seen under such favourable circumstances, shining steadily as a minus-magnitude light low in a dark twilight sky, many first-time Mercury observers wonder why they had ever imagined that the planet was difficult to see.

At favourable (springtime) evening elongations in mid-northern latitudes, Mercury first becomes noticeable with the unaided eye around half an hour after sunset, at the start of civil twilight when the Sun reaches an altitude of 6° below the western horizon. Depending on the flatness of the horizon and the clarity of the atmosphere, the planet can remain visible in a darkening twilight sky for up to an hour until eventually fading as its light is extinguished near the horizon. During favourable morning apparitions in the northern hemisphere it is possible to pick up Mercury with the unaided eye within half an hour of its rising, and it can be followed without optical aid for more than an hour before fading from view in brightening skies at the end of civil twilight. At Mercury's most favourable eastern and western elongations for southern hemisphere observers it is possible to view the planet for a longer period of time – for about an hour and a half – between the start of civil twilight, through astronomical twilight and beyond.

At its brightest, Mercury shines at around magnitude −1.3, slightly fainter than Sirius, the sky's brightest star at magnitude −1.4. The planet's low altitude and distortion within its light path by air currents often makes it appear to scintillate. Mercury is often described as having a rosy or pinkish hue, but the subtle colour (most noticeable through binoculars) is another effect produced by the planet's low altitude.

Binocular Observations

While regular binoculars won't reveal Mercury's phase, they provide a means by which the planet can be located in twilight skies some time before the unaided eye. A steadily-held pair of binoculars will nicely show close planetary conjunctions and

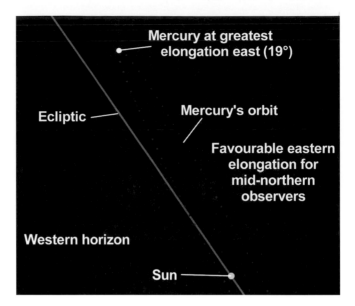

A favourable eastern elongation for mid-northern hemisphere observers, evening, northern hemisphere spring. Peter Grego.

appulses with the Moon and the stars. A word of warning – attempts to locate Mercury through binoculars by sweeping the sky near the Sun must never be made when the Sun is in full view; even when low down the Sun is bright enough to cause permanent and irreversible eye damage if its light is inadvertently directed into the eyes through any optical equipment.

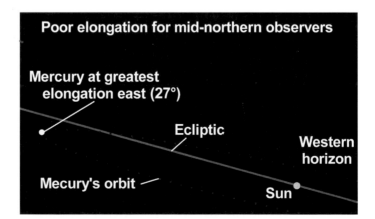

A poor eastern elongation for mid-northern hemisphere observers, evening, northern hemisphere autumn. Peter Grego.

Mercury's Phases and its Appearance Through the Telescope

Mercury can be observed telescopically for a maximum of around five weeks during favourable apparitions. Being an inner planet, Mercury displays a sequence of phases when viewed through even a small telescope used at a fairly high magnification.

At superior conjunction Mercury is a fully illuminated disk – far too close to the Sun for regular amateur observation – and depending on its distance from the Earth on these occasions its apparent diameter varies between 4.7 and 5.1 arcseconds. Following superior conjunction, as Mercury pulls away from the Sun in the evening skies, its phase decreases, and as its distance from the Earth decreases its apparent angular diameter grows. Mercury reaches its brightest magnitude a week or two before greatest elongation east or following greatest elongation west when it is a wide gibbous phase.

At greatest elongation east, Mercury has reached a half illuminated phase with an apparent disk diameter of between 6.7 and 8.2 arcseconds in diameter. The planet subsequently becomes a crescent, slowly growing in size by a couple of arcseconds until it is lost in the evening twilight skies as it draws closer to the Sun and its inferior conjunction. At inferior conjunction Mercury's Earth-facing hemisphere is completely unilluminated, and completely unobservable except on those rare occasions when it can be discerned as a silhouette of between 10 and 12 arcseconds diameter during transits across the Sun's face.

Following inferior conjunction the planet moves to the west of the Sun, venturing out into the predawn skies. During the course of the next few weeks it displays a

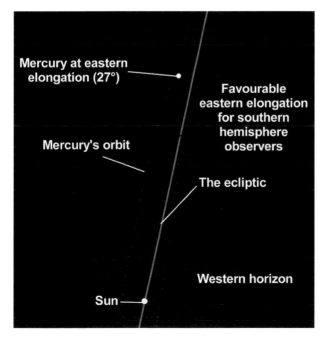

A good eastern elongation for southern hemisphere observers, southern spring evening. Peter Grego.

'waxing' phase sequence from crescent, through half illuminated to gibbous phase. During favourable western elongations Mercury first becomes observable when it is around 15° west of the Sun, rising more than an hour before sunrise and presenting a narrow crescent around 10 arcseconds in apparent diameter. Greatest elongation west takes place around a fortnight later, the planet being a half-illuminated phase (with the same range of apparent diameters as at greatest eastern elongation). As Mercury moves eastwards, drawing back towards the Sun, its apparent diameter diminishes yet further and its phase increases to gibbous.

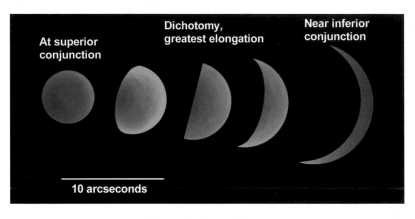

Mercury's phases. Grego.

Long before its surface had been imaged up close, much had been deduced about the true nature of Mercury's surface from visual observations alone. Mercury's albedo (a measurement of the amount of sunlight it reflects) is 0.1 – the lowest albedo of all the Solar System's major planets, equivalent to the overall albedo of the Moon but offset because of the planet's closeness to the Sun. Mercury's light curve, from its bright limb to the terminator (the division between the planet's illuminated and unilluminated hemispheres) is also similar to that of the Moon. Some observers have compared a good telescopic view of Mercury under excellent seeing conditions with the unaided eye view of the Moon. However, regional albedo variations are much less pronounced on Mercury than on the Moon; while the Moon has a distinct, clearly defined patchwork of dark maria, brighter highland areas and broad ejecta splotches, Mercury's overall shading is far more subtle. When viewed at greatest elongation around dichotomy, the planet's terminator shading is similar to that visible on our own first quarter Moon when seen with the unaided eye during daylight, or low down in the evening when its brightness is reduced by haze.

Using large instruments under the most favourable conditions, a few sharp-eyed observers (some of them professional astronomers) have discerned craters along Mercury's terminator (the division between the planet's illuminated and unillumi-nated hemispheres) where the low angle of illumination causes topographic features to cast shadows. While such observations may only be made under exceptional circumstances, the topographically rough nature of airless Mercury is eluded to by the Moon-like degree of terminator shading. Sometimes one or other (or both) or Mercury's cusps can appear somewhat blunted or muted in tone, due to the presence of heavily cratered regions near the terminator in the polar regions.

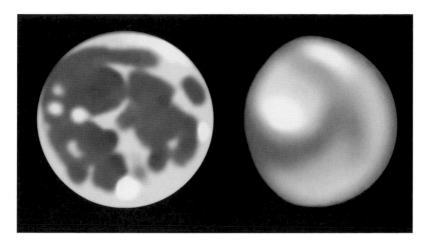

Estimated tonal comparison between an actual unaided eye view of the Moon compared with a telescopic view of Mercury under excellent seeing conditions. Grego.

Under good observing conditions, a 100 mm telescope at a magnification of 100× is sufficient to discern shading along Mercury's terminator when the planet is around its greatest elongation. To the untrained eye, the Mecurian disk displays little in the way of obvious markings such as those found on Mars, for example.

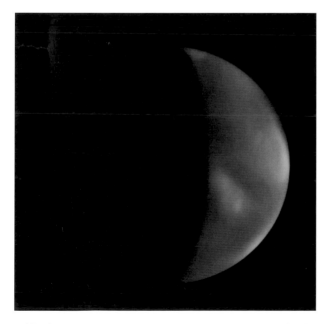

Mercury, observed by the author on 3 May 2002 using a 127 mm MCT. The dark area near the centre of the crescent is the dusky Solitudo Criophori, while the brighter area nearby is Pieria. The north polar cusp was noted to be slightly brighter than the south. Peter Grego.

Mercury, observed by the author on 3 April 1991 using a 100 mm refractor. The south cusp appeared more strongly defined than the north. The dark area in the north is Solitudo Aphrodites. Peter Grego.

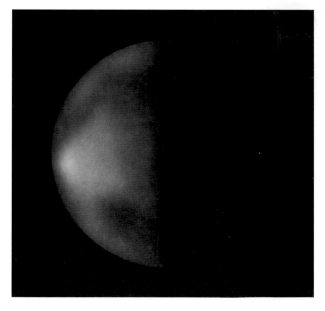

Mercury, observed by the author on 8 September 1991 using a 150 mm reflector. The dark area on the lower terminator is Solitudo Martis, while the brighter area towards the limb is Phaethontias. Peter Grego.

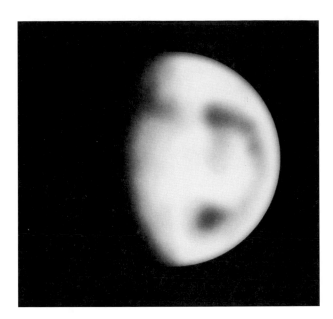

Mercury, observed by the author on 12 March 1998. Solitudo Admetei is the dark area along the northern terminator. Peter Grego.

However, experienced observers usually have little difficulty in discerning albedo shadings on the planet's illuminated face. These markings correspond with the reflectivity of large tracts of Mercury's surface, yet on casual inspection they bear little resemblance to the planet's known topography. On the Moon, for example, the dark areas correspond with large, smooth lava-filled marial basins, while the brighter areas are generally heavily crated highland regions. Although the markings on Mercury are more subtle, they are plain enough for the IAU to have sanctioned

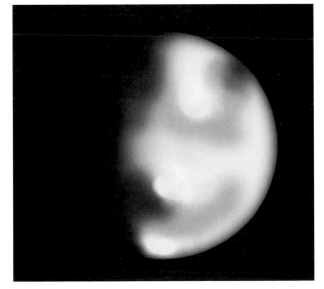

Mercury, observed by the author on 17 May 2007 using a 200 mm SCT ×250. Peter Grego. The dark area on the northern terminator is Solitudo Admetei, while the dark area in the south is Solitudo Martis. The dusky feature towards the northern limb is in the Apollonia region. Peter Grego.

an official albedo map and associated albedo nomenclature to which the amateur observer may refer.

Visually, Mercury is a planet of two distinct sides. From longitudes 90 to 270° west, in both northern and southern hemispheres, there are a number of large dark patches, notably Solitudo Phoenicis and Solitudo Neptuni in the north, and Solitudo Atlantis and Solitudo Martis in the south. Several brighter tracts run between these areas, including Pleias Gallia to the east of Solitudo Neptuni and the equatorial Phaethontias. The planet's other half, from 270 to 90° west, is composed of more subtle shadings and includes the extensive bright regions of Aurora and Apollonia in the north, Cyllene and Solitudo Hermae Trismegisti in the south.

When making pencil sketches, a 50 mm circular template is commonly used to depict Mercury. The planet's phase is first drawn in, and any detail is then added. Bold, sharply defined features are unlikely to be visible; shading is likely to be faint and indistinct, so be easy on the graphite and exaggerate only if necessary. Brighter areas may be outlined by dotted lines and/or by using an eraser. Lightly draw in any terminator shading and any apparent irregularities that may be visible along the terminator – these irregularities may represent albedo markings or even shadows being cast by topographic features. Intensity estimates are made on a scale of 0

Mercurian albedo features

1. Apollonia	45N, 315W	Extensive bright albedo region.
2. Aurora	45N, 90W	Bright albedo region.
3. Australia	73S, 0W	Albedo feature (not on chart).
4. Borea	75N, 0W	Albedo feature (not on chart).
5. Caduceata	45N 135W	Dusky albedo region.
6. Cyllene	41S, 270W	Bright albedo region.
7. Heliocaminus	40N, 170W	Dusky albedo tract.
8. Hesperis	45S, 355W	Bright albedo feature.
9. Liguria	45N, 225W	Bright albedo tract.
10. Pentas	5N, 310W	Bright albedo feature.
11. Phaethontias	0N, 167W	Bright albedo tract.
12. Pieria	0N, 270W	Bright albedo feature.
13. Pleias Gallia	25N, 130W	Bright albedo feature.
14. Sinus Argiphontae	10S, 335W	Dusky albedo feature.
15. Solitudo Admetei	55N, 90W	Albedo feature.
16. Solitudo Alarum	15S, 290W	Dusky albedo feature.
17. Solitudo Aphrodites	25N, 290W	Dusky albedo feature.
18. Solitudo Atlantis	35S, 210W	Dark albedo feature.
19. Solitudo Criophori	0N, 230W	Dusky albedo feature.
20. Solitudo Helii	10S, 180W	Dark albedo feature.
21. Solitudo Hermae Trismegisti	45S, 45W	Large bright albedo region.
22. Solitudo Horarum	25N, 115W	Dusky albedo feature.
23. Solitudo Iovis	0N, 0W	Dark albedo feature.
24. Solitudo Lycaonis	0N, 107W	Albedo feature.
25. Solitudo Maiae	15S, 155W	Dark albedo feature.
26. Solitudo Martis	35S, 100W	Dark albedo feature.
27. Solitudo Neptuni	41S, 225W	Dusky albedo region.
28. Solitudo Phoenicis	25N, 225W	Large dark albedo feature.
29. Solitudo Promethei	45S, 143W	Dark albedo feature.
30. Tricrena	0N, 36W	Dusky albedo feature.

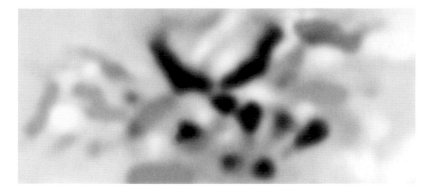

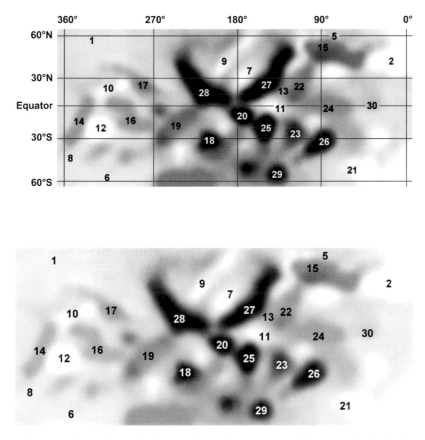

Map of Mercury based on official IAU chart, showing most of the named albedo features (numbering key as feature list above). Grego.

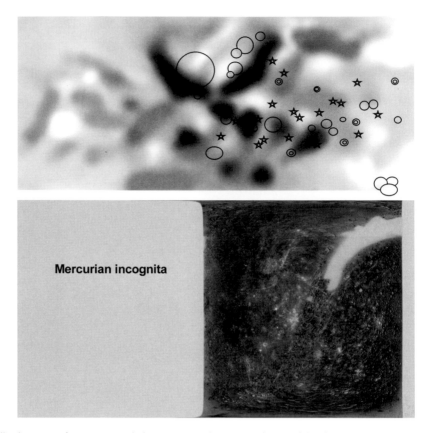

Albedo map of Mercury, with large impact basins (circles) and bright impact craters (stars) superimposed, compared with a topographic map of the planet. The largest known basin is Caloris, while the furthest impact crater to the right (east) is Kuiper. The area of Mercurian incognita is the half the planet that remains to be imaged in detail by spaceprobes. As will be seen by the comparison, the albedo markings give us few clues about the topography of the incognita side – there are likely to be numerous large basins and bright impact craters, but it does seem certain that there are no dark maria-type plains.

to 5, where 0 represents the most brilliant features (like impact craters and their bright ejecta) and 5 the most prominent dark shadings.

Colour filters are capable of visually enhancing features on Mercury. When observing in brighter twilight conditions, an orange, light red, red or deep red filter (Wratten 21, 23a, 25 and 29 or equivalents, respectively) will enhance the contrast between the planet and the background sky and improve the quality of the seeing. Of these, the orange filter offers the brightest image but is least effective, while the deep red is best used on a large instrument under good conditions. Some observers have found violet, green and light blue filters (Wratten 47, 68 and 80a or equivalents, respectively) effective at bringing out the darker shadings.

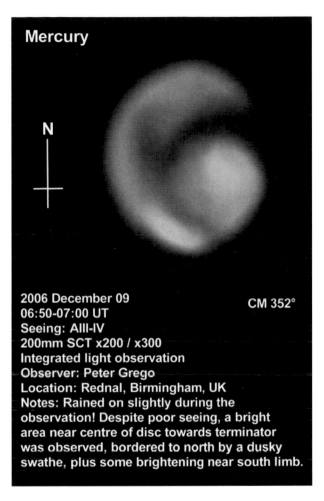

Mercury

N

2006 December 09
06:50-07:00 UT
Seeing: AIII-IV
200mm SCT x200 / x300
Integrated light observation
Observer: Peter Grego
Location: Rednal, Birmingham, UK
Notes: Rained on slightly during the
observation! Despite poor seeing, a bright
area near centre of disc towards terminator
was observed, bordered to north by a dusky
swathe, plus some brightening near south limb.

CM 352°

An example of a finished observational drawing of Mercury. Observation made on 9 December 2006. Peter Grego.

Elongations of Mercury 2009–2019

Elongation	Date	Distance from Sun
Eastern	2009 Jan 4	19.2°
Western	2009 Feb 13	26.3°
Eastern	2009 Apr 26	20.2°
Western	2009 Jun 13	23.2°
Eastern	2009 Aug 24	27.3°
Western	2009 Oct 6	17.2°
Eastern	2009 Dec 18	20.2°
Western	2010 Jan 27	24.2°
Eastern	2010 Apr 8	19.2°

(Continued)

Elongations of Mercury 2009–2019 (Continued)

Elongation	Date	Distance from Sun
Western	2010 May 26	25.3°
Eastern	2010 Aug 7	27.3°
Western	2010 Sep 19	17.2°
Eastern	2010 Dec 1	21.2°
Western	2011 Jan 9	23.2°
Eastern	2011 Mar 23	18.2°
Western	2011 May 7	26.3°
Eastern	2011 Jul 20	26.3°
Western	2011 Sep 3	18.2°
Eastern	2011 Nov 14	22.2°
Western	2011 Dec 23	21.2°
Eastern	2012 Mar 5	18.2°
Western	2012 Apr 18	27.3°
Eastern	2012 Jul 1	25.3°
Western	2012 Aug 16	18.2°
Eastern	2012 Oct 26	24.2°
Western	2012 Dec 4	20.2°
Eastern	2013 Feb 16	18.2°
Western	2013 Mar 31	27.3°
Eastern	2013 Jun 12	24.2°
Western	2013 Jul 30	19.2°
Eastern	2013 Oct 9	25.3°
Western	2013 Nov 18	19.2°
Eastern	2014 Jan 31	18.2°
Western	2014 Mar 14	27.3°
Eastern	2014 May 25	22.2°
Western	2014 Jul 12	20.2°
Eastern	2014 Sep 21	26.3°
Western	2014 Nov 1	18.2°
Eastern	2015 Jan 14	18.2°
Western	2015 Feb 24	26.3°
Eastern	2015 May 7	21.2°
Western	2015 Jun 24	22.2°
Eastern	2015 Sep 4	27.3°
Western	2015 Oct 16	18.2°
Eastern	2015 Dec 29	19.2°
Western	2016 Feb 7	25.3°
Eastern	2016 Apr 18	19.2°
Western	2016 Jun 5	24.2°
Eastern	2016 Aug 16	27.3°
Western	2016 Sep 28	17.2°
Eastern	2016 Dec 11	20.2°
Western	2017 Jan 19	24.2°
Eastern	2017 Apr 1	19.2°
Western	2017 May 17	25.3°
Eastern	2017 Jul 30	27.3°
Western	2017 Sep 12	17.2°
Eastern	2017 Nov 24	22.2°
Western	2018 Jan 1	22.2°
Eastern	2018 Mar 15	18.2°

(Continued)

Elongations of Mercury 2009–2019 (*Continued*)

Elongation	Date	Distance from Sun
Western	2018 Apr 29	27.3°
Eastern	2018 Jul 12	26.3°
Western	2018 Aug 26	18.2°
Eastern	2018 Nov 6	23.2°
Western	2018 Dec 15	21.2°
Eastern	2019 Feb 27	18.2°
Western	2019 Apr 11	27.3°
Eastern	2019 Jun 23	25.3°
Western	2019 Aug 9	19.2°
Eastern	2019 Oct 20	24.2°
Western	2019 Nov 28	20.2°

The Stroboscope Effect

As we have noted, not all elongations of Mercury are favourable – the best elongations occur when the planet appears highest in the sky after sunset or before sunrise, and this depends on the observer's location on the Earth and the time of year in which the planet is observed. Because of this, observers restricted to picking and choosing the best elongations are presented with a somewhat biased view of the planet's surface. Known as the stroboscope effect, an observer will be presented with much the same longitude during a number of successive favourable elongations. This effect caused observers in the past to wrongly deduce that Mercury was in a captured rotation, keeping the same face turned towards the Sun throughout its orbit. The stroboscope effect only works over a two or three year period – eventually, orbital dynamics cancel it out, allowing a dedicated observer to view the entire planet over a longer period of time, even if the observations are made only during favourable elongations.

Transits of Mercury

Mercury's synodic period – the interval between inferior conjunctions – is around 116 days. However, its orbit is inclined by about 7° to the plane of the ecliptic, so Mercury usually passes far to the north or south of the Sun at inferior conjunction. Transits take place when Mercury moves directly across the line joining the Sun and the Earth. No fewer than fourteen transits of Mercury took place during the 20th Century. The combined orbital circumstances of Mercury and the Earth cause transits to happen only during the months of May and November. May transits recur every 13 and 33 years, while November transits recur at intervals of 7, 13, and 33 years. Since Mercury is considerably closer to the Earth during May transits, it appears as a black disc some 12 arcseconds in apparent diameter, while its disc is just 10 arcseconds across when viewed during November transits. It is, however, far too small to be seen with a protected naked eye, and viewing it safely requires a telescope equipped with a full aperture solar filter or by projecting the image onto a shielded white card.

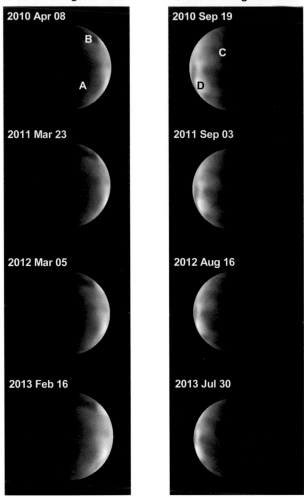

Favourable elongations from northern latitudes

Greatest elongation east	Greatest elongation west
2010 Apr 08	2010 Sep 19
2011 Mar 23	2011 Sep 03
2012 Mar 05	2012 Aug 16
2013 Feb 16	2013 Jul 30

Showing the aspect of Mercury at each of the most favourable elongations east (left column) and west (right column) as viewed from the northern hemisphere between 2010 and 2017. Features within the unilluminated hemisphere are shown for clarity. The stroboscope effect can clearly be seen in successive elongations. (*Figure continued on next page*)

Key:
A *Solitudo Martis*
B *Solitudo Admetei*
C *Solitudo Aphrodites*
D *Pentas*
E *Solitudo Criophori*
F *Solitudo Neptuni*

Favourable elongations from northern latitudes

Greatest elongation east	Greatest elongation west

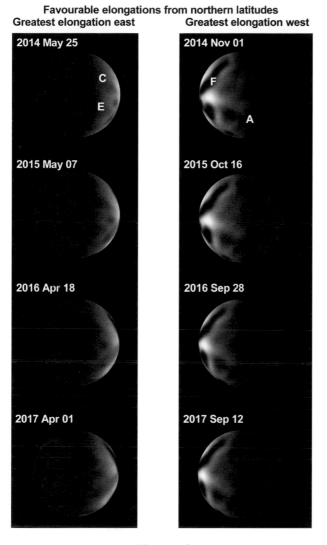

From its first appearance as a tiny indentation on the Sun's limb (after first contact), Mercury can take as little as two minutes to enter fully upon the Sun (second contact), depending on the angle at which it approaches the solar limb. In the past, some observers have claimed a 'black drop' effect, where the planet briefly appears to remain attached to the Sun's limb after second contact. Such an effect is caused by a combination of atmospheric seeing and the instrument's own optics. Mercury has no atmosphere to speak of, so none of the fascinating phenomena observable during a transit of Venus, which has a heavy, highly refractive atmosphere, are visible. Depending on the size of the chord traced by the planet's movement against the Sun's disk, transits of Mercury can last from a few minutes (for a grazing transit at the Sun's limb) to more than four hours.

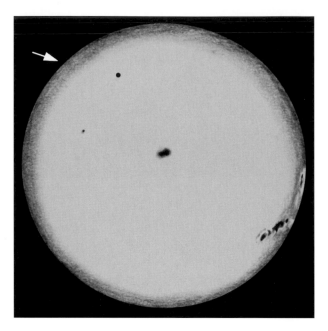

The transit of Mercury of 7 May 2003, observed by Peter Grego.

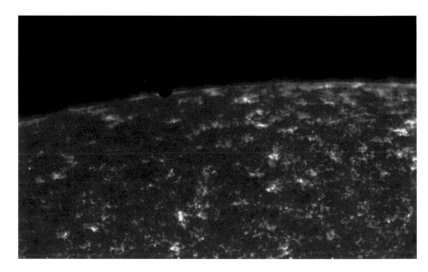

The 2003 transit of Mercury imaged by NASA's SOHO observatory in H-alpha light. NASA / SOHO.

Future transits of Mercury

Date	Time (UT, mid-transit)	Duration
2016 May 09	14:57	07 h 30 m
2019 Nov 11	15:20	05 h 29 m
2032 Nov 13	08:58	04 h 26 m

Occultations

Occultations of Mercury by the Moon, along with occultations of bright stars by Mercury, are fascinating to observe although very rare. Only two lunar occultations of the planet are visible in the period to 2019 – these take place on 14 November 2012 and 26 June 2014.

Bright 21st Century stellar and planetary occultations involving Mercury

Date	Time	Event
2052 Nov 10	07:20	Mercury occults Zuben-el-genubi
2067 Jul 15	11:56	Mercury occults Neptune
2079 Aug 11	01:30	Mercury occults Mars
2088 Oct 27	13:43	Mercury occults Jupiter
2094 Apr 7	10:48	Mercury occults Jupiter

Chapter 7

Observing Venus

Brilliant Morning and Evening Star

Venus is the second major planet from the Sun. Known as a 'wandering star' since humans first looked up at the sky with curiosity, it regularly appears as a brilliant morning or evening 'star', outshining every other planet and star with a dazzling maximum magnitude of −4.6. With a brilliance surpassed only by the Sun and the Moon (and occasional bright meteors) Venus is so bright that it is capable of casting clearly discernable shadows when its light alone (apart from the stars) is enabled to illuminate a scene. Venus' dazzling presence in the evening or morning skies often attracts the attention of non-astronomers, some of whom imagine it to be the lights of a relatively nearby aerial object. Venus shines with a steady, brilliant white hue, and even when it is fairly low near the horizon it may not show scintillation due to atmospheric turbulence.

Apparitions

To describe Venus' typical apparitions, let's examine Venus as it makes an entire lap of the Sun (as seen from the Earth) beginning at superior conjunction, through eastern elongation, inferior conjunction, western elongation and ending with superior conjunction once more.

At superior conjunction Venus is furthest from the Earth, on the far side of its orbit and some 525 million kilometres distant. It appears as a fully illuminated disk some 9.7 arcseconds in diameter, but it is far too close to the Sun's glare for practical amateur observation, and indeed the planet occasionally moves directly behind the Sun itself at superior conjunction. Perspective effects mean that the average apparent angular distance between Venus and the Sun is far closer at superior conjunction than at inferior conjunction; so, while it is not possible to observe Venus near superior conjunction it is often possible for the experienced observer to view the planet around inferior conjunction when it is a narrow crescent phase (see below).

Within two months after superior conjunction the planet has edged sufficiently east of the Sun to become visible with the unaided eye low in the evening twilight skies, shining at around magnitude −3.8. At this time its apparent diameter is around 10 arcseconds, barely larger than when at superior conjunction; its broad gibbous phase is large enough to discern through a small telescope, say a 60 mm refractor, at a magnification of ×100. The planet continues to move

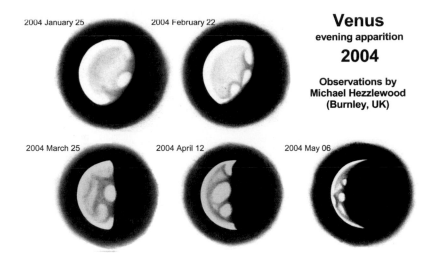

Venus' evening apparition in 2004, observed by Michael Hezzlewood on January 25, February 22, March 25, April 12 and May 6 using a 128mm APO refractor and 102mm refractor (latter three).

eastwards, slowly growing in apparent diameter while its gibbous phase continues to decrease.

Venus takes around seven months to reach its maximum eastern elongation, between 45° and 47° from the Sun – nearly twice the angular distance of the best elongations of Mercury and generous enough for the planet to be seen against a truly dark evening sky, several hours after sunset during favourable apparitions. At eastern elongation, Venus shines at around magnitude –4.2 and presents a half phase disk some 25 arcseconds in diameter.

Maximum eastern elongations of Venus are most favourable from mid-northern latitudes during the spring, when the planet can be as high as 40° above the western horizon at sunset and observable for more than four hours in darkening evening skies. For observers in northern temperate regions, unfavourable eastern elongations take place during the autumn, when the angle made by Venus, the setting Sun and the horizon is at its smallest. On these occasions, Venus can be less than 10° above the southwestern horizon at sunset. With such a low altitude, the telescopic image of Venus is liable to suffer greatly from the effects of atmospheric turbulence.

Venus' journey back westwards towards the Sun is made at a much faster pace than its outward leg. It takes around ten weeks or so to move from maximum eastern elongation to inferior conjunction. Although the planet's phase becomes an increasingly narrower crescent following eastern dichotomy, its apparent diameter continues to grow. Venus reaches magnitude –4.6, its maximum brilliance in the evening skies, around five weeks before inferior conjunction when at an elongation of around 39° east of the Sun. Telescopically it presents a crescent phase, a little more than 25 percent illuminated, with an apparent diameter of some 40 arcseconds.

Inferior conjunction takes place when the planet lies between the Sun and the Earth. Precise alignments between the three objects are even rarer than with

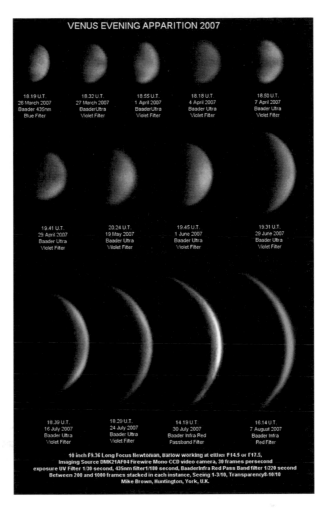

VENUS EVENING APPARITION 2007

18.19 U.T. 26 March 2007 Baader 435nm Blue Filter	18.32 U.T. 27 March 2007 BaaderUltra Violet Filter	18.55 U.T. 1 April 2007 BaaderUltra Violet Filter	18.18 U.T. 4 April 2007 Baader Ultra Violet Filter	18.50 U.T. 7 April 2007 Baader Ultra Violet Filter

19.41 U.T. 29 April 2007 Baader Ultra Violet Filter	20.24 U.T. 19 May 2007 Baader Ultra Violet Filter	19.45 U.T. 1 June 2007 Baader Ultra Violet Filter	19.31 U.T. 29 June 2007 Baader Ultra Violet Filter

18.39 U.T. 16 July 2007 Baader Ultra Violet Filter	18.29 U.T. 24 July 2007 Baader Ultra Violet Filter	14.19 U.T. 30 July 2007 Baader Infra Red Passband Filter	16.14 U.T. 7 August 2007 Baader Infra Red Filter

10 inch F9.36 Long Focus Newtonian, Barlow working at either F14.5 or F17.5,
Imaging Source DMK21AF04 Firewire Mono CCD video camera, 30 frames per second
exposure UV Filter 1/30 second, 435nm filter1/180 second, BaaderInfra Red Pass Band filter 1/220 second
Between 200 and 1000 frames stacked in each instance, Seeing 1-3/10, Transparency8-10/10
Mike Brown, Huntington, York, U.K.

A full sequence of Venus' phases during the evening apparition of 2007, captured on CCD by Mike Brown using a 250mm Newtonian.

Mercury; and transits of Venus across the Sun's face are one of astronomy's most eagerly anticipated events (see below). Most often, Venus passes some considerable distance north or south of the Sun.

Rapidly moving to the west of the Sun, Venus quickly establishes itself in the morning skies before dawn. During favourable apparitions it can be picked up with the unaided eye within a few weeks of inferior conjunction, low in the eastern sky prior to sunrise. Maximum brightness is reached around five weeks into the apparition when Venus is around 39° east of the Sun, a crescent phase more than a quarter illuminated and shining at magnitude −4.3. Around ten weeks after inferior conjunction Venus has reached its maximum western elongation from the Sun, between 45° and 47°; it is then at half phase with an apparent angular diameter of around 25 arcseconds. Observers in north temperate latitudes find that maximum western elongations are at their most favourable when they occur during the autumn, when Venus rises above the eastern horizon almost five hours before the Sun. Springtime maximum western elongations are the poorest from an observational point of view, with Venus barely visible above the south-eastern horizon before dawn. Following western elongation it takes around seven

(a)

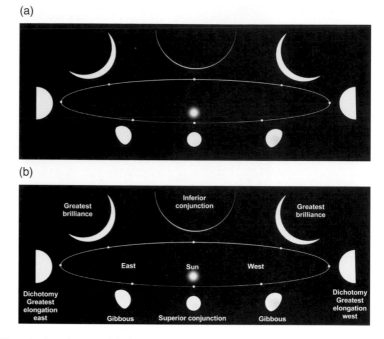

(b)

Venus' orbit, showing the planet's phases and apparent diameter (to scale). Grego.

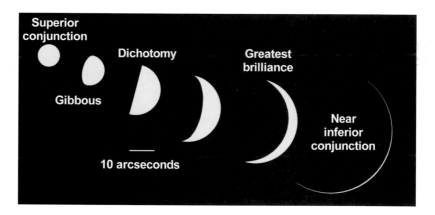

The phases of Venus. Peter Grego.

months to return to superior conjunction once more during which time the planet broadens out into a gibbous phase, gradually becoming smaller during its journey eastwards towards the Sun. Periods between superior conjunctions average 584 days.

Elongations of Venus 2009 to 2019

Elongation	Date	Distance from Sun
Eastern	2009 Jan 14	47.5°
Western	2009 Jun 5	45.5°
Eastern	2010 Aug 20	46.5°
Western	2011 Jan 8	47.5°
Eastern	2012 Mar 27	46.5°
Western	2012 Aug 15	45.5°
Eastern	2013 Nov 1	47.5°
Western	2014 Mar 22	46.5°
Eastern	2015 Jun 6	45.5°
Western	2015 Oct 26	46.5°
Eastern	2017 Jan 12	47.5°
Western	2017 Jun 3	45.5°
Eastern	2018 Aug 17	45.5°
Western	2019 Jan 6	47.5°

Inferior conjunctions of Venus 2009 to 2019

Date	Distance from Sun
2009 Mar 28	8°N
2010 Oct 29	6°S
2012 Jun 6	Transit
2014 Jan 11	5°N
2015 Aug 15	8°S
2017 Mar 25	8°N
2018 Oct 26	6°S

Daytime Views with the Unaided Eye

One usually failsafe means of seeing Venus (and Mercury, too, when of sufficient brightness) during the day is to view one of the most awesome sights in nature, a total eclipse of the Sun. For a few brief moments – sadly never more than 7 minutes and 40 seconds – the solar disk is completely hidden by the Moon. In the strange darkness of totality the brighter stars and planets become easily visible; some of them are perhaps near enough to the Sun to be embedded in the pearly streamers of the corona, the Sun's outer atmosphere. The circumstances for viewing Mercury and Venus during forthcoming total solar eclipses are as follows:

Date of eclipse	Mercury location & magnitude	Venus location & magnitude
2009 Jul 22	9°E, −1.2	40°W, −4.0
2010 Jul 11	15°E, −0.8	42°E, −4.1
2012 Nov 13	9°E, 2.9	32°W, −4.0
2015 Mar 20	18°W, −0.4	34°E, −4.0
2016 Mar 09	12°W, −0.7	23°W, −3.9
2017 Aug 21	10°E, 3.4	34°W, −4.0
2019 Jul 02	23°E, 1.3	12°W, −3.9

Total eclipses aside, the easiest method of viewing Venus in full daylight without optical aid is to find it before dawn and continue to keep it in sight it for as long as possible after sunrise. During favourable elongations, given a transparent sky, it is possible to keep Venus in view by keeping the Sun obscured by foreground objects such as a nearby building, and adjusting the view so that Venus is kept in line with a reference object, such as the branch of a tree.

Venus can be located from scratch in a clear, haze-free sky during full daylight with the unaided eye. Unfortunately, such circumstances present themselves rather infrequently in urban locations that are prone to smog or in parts of the world that regularly find their skies smothered with aircraft contrails. To have the best chance of locating Venus in daylight with the naked eye, the Sun must be hidden from direct view by using a convenient local object, such as the edge of a roof or the side of a building; alternatively, choose a time when Venus is high in the sky and the Sun is obscured by a distant landscape feature such as a mountain. Naturally, the attempt ought to be made knowing the approximate location of Venus and its angular distance from the Sun; this is where even the most basic planetarium program for your computer will prove its worth.

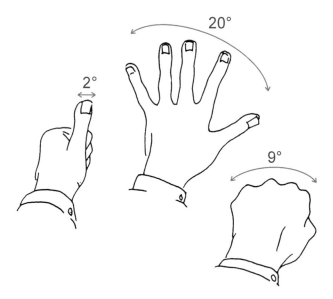

Angular distances in the sky can be estimated using the hands. At greatest elongations Venus is more than two fully outstretched hands' widths away from the Sun.

While gazing at a bland area of sky, the eyes have no reference object in the far distance upon which to focus, which means that attempting to locate a faint pinprick of light in a bright sky can be something of a challenge. One helpful technique is to first scan the far horizon and then quickly turn one's gaze to the area of sky in which Venus is thought to be located; since the eyes will retain their distance focus for a short while, the observer's chances of finding the planet are increased.

Binocular Observations

Binoculars are great instruments for taking in wide angle views of Venus and its surrounding stars, observing close planetary appulses and close approaches to the Moon. A steadily-held, mounted pair of large high magnification binoculars, such as a pair of 15 × 70s, is just about capable of revealing the planet's crescent phase when it is larger than 40 arcseconds in apparent diameter.

Under no circumstances should attempts be made to visually sweep for Venus through binoculars when the Sun is nearby and in direct view; the briefest of exposures to magnified sunlight is capable of damaging the eye's sensitive retina and ruining an observer's eyesight permanently. Attempts to find Venus in the day using binoculars are best made when the skies are clear and blue, and distant objects can be seen clearly on the horizon. Pollution and atmospheric haze will reduce Venus' brightness and make it harder to find. One of the best celestial stepping stones for finding Venus is the Moon, an object which can usually be found in the daytime sky without any difficulty when it is above the horizon. Every lunar month, the Moon passes within a hand's span of Venus, often coming within a few lunar diameters' distance. A computer program or astronomical almanac enables to observer to find out exactly where Venus is in relation to the Moon at a given time during a daytime appulse, making it easier to locate by just sweeping near to the Moon.

The appulse of Venus and Saturn, 1° apart, as seen through high power binoculars on 30 June 2007. Grego.

Telescopic Observation

Through a telescope, the sheer brilliance of Venus can dazzle the eye so much that it can initially be difficult to perceive the planet's phase, let alone any vague atmospheric detail that might be present. Venus' glare is capable of producing some pretty spectacular, though wholly unwanted, special effects. Budget refractors may show Venus surrounded by coloured fringes, caused by chromatic aberration, and certain types of eyepiece may produce internal reflections and ghosting. Another unwanted effect called irradiation, visible through even the best optics, causes blurring between areas of greatly differing brightness. Irradiation can be produced physiologically by the observer's eye, or as a result of atmospheric turbulence. One way of reducing Venus' glare is to observe when it is high in a bright twilight sky, or even during the daytime.

Locating Venus in Daytime Through a Telescope

Venus can be telescopically observed during the daytime for much of its orbit. Only around a fortnight or so on either side of superior conjunction does Venus completely hide itself from amateur scrutiny because of its sheer proximity to the Sun – it is never more than 2° away from the Sun at superior conjunction. Venus can be as far as 8° north or south of the Sun at inferior conjunction, making it possible to locate the planet through a telescope.

To prevent any possibility of eye damage when sweeping for Venus in the daytime through a telescope, extreme care must be taken not to accidentally sweep across the unfiltered image of the Sun.

Using a properly set up computerised telescope, Venus can be picked up in broad daylight without a great deal of difficulty. The traditional way of finding Venus during the daytime is to use a well-aligned equatorial instrument with accurate setting circles. First, the Sun is centred in the telescope. To do this, most observers keep the covers on their main instrument, finderscope and eyepieces, and align the instrument by the shadow method or any safe device that allows a good approximation of pointing sunwards. Having noted the co-ordinates of the Sun (by consulting an ephemeris or computer program) for the time of the observation, these co-ordinates are transferred to the telescope's RA and Declination circles and the drive is engaged. The observer is now ready to offset the telescope to the appropriate amount, remove the covers and begin sweeping through the eyepiece.

When sweeping for Venus, one very helpful useful accessory to use is a low-magnification, wide field eyepiece with a small opaque pointer fixed in its optical plane. The eyepiece is first calibrated by focusing on a distant object; while sweeping bland, featureless tracts of sky, the observer's focus is kept pin-sharp by the presence of the internal pointer. Additionally, an orange filter will increase the contrast between Venus and the background sky. Once Venus is found, higher magnifications may be used to view the planet; orange and red filters help boost contrast and improve seeing, allowing more subtle cloud detail to be seen. When near inferior conjunction Venus appears as a wafer-thin crescent with a surprisingly large apparent diameter of around 60 arcseconds, 1/30 the apparent diameter of the Moon. At this stage, its cusps extend for some distance around the limb, a phenomenon caused by the scattering of sunlight within Venus' atmosphere. Around inferior conjunction, the cusps actually extend so far as to meet up on the other side of the planet, producing a remarkable annular phase.

Nocturnal Observations of Venus

Most observers restrict their views of Venus to times of twilight or night, mainly because of the advantage of convenience – the planet is so bright that no time at all is spent in trying to locate it. One major disadvantage to nocturnal observation of Venus is its sheer brilliance, a glare that can easily mask subtle cloud detail. Variable density polarizing filters can reduce this glare and enhance the visibility of faint markings. Another downside to nocturnal observation is the planet's relatively low altitude in the night sky – even at its most favourable elongations, Venus is never higher than 30º high in an astronomically dark sky.

It's generally recommended that when possible Venus should be observed when it is higher than 20º above the horizon, when it is clear of much of the murk and turbulence within the Earth's atmosphere. For observers in northern or southern temperate regions, some apparitions of the planet are so unfavourable as to keep Venus beneath 20º in altitude during darkness and twilight, even when it is at its maximum elongation from the Sun. From mid-northern latitudes, autumnal eastern elongations and springtime western elongations are unfavourable, while for mid-southern latitudes autumn western elongations and spring eastern elongations are unfavourable.

Venus' Cloud Patterns

Constant high-speed winds in the order of 360 km/hour in Venus' upper atmosphere drive the clouds around the planet; a decrease in wind speed from the planet's equator to the poles causes the appearance of a uniform atmospheric rotation with a period of four or five days. Venus' upper atmospheric rotation can be observed through a telescope as small as 100 mm. During eastern elongations of Venus, cloud features appear to move from the bright limb to the terminator; during western elongations, cloud features appear at the terminator and are carried across the disc to the bright limb. If careful observations are made under good conditions, the drifting atmospheric patterns can be revealed during the course of a few hours.

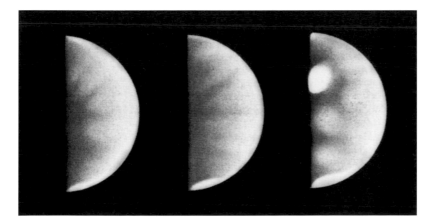

Changes in the appearance of Venus near dichotomy on 1991 November 3, 4 and 5, clearly showing polar hoods, dusky features and (in the latter) a bright region on the terminator. Observations made with a 115mm refractor ×186, by Richard Baum.

(a)

Independent visual observations of Venus compared

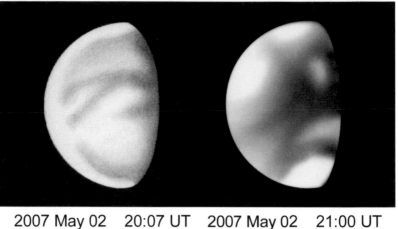

2007 May 02   20:07 UT   2007 May 02   21:00 UT
90mm MCT x80 x160   127mm MCT x200
Nigel Longshaw   Peter Grego
(Chadderton, Oldham, UK) (Rednal, Birmingham, UK)

(b)

Independent visual observations of Venus compared

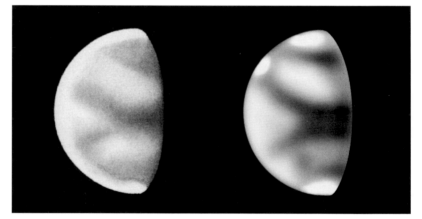

2007 May 07   19:30 UT   2007 May 07   20:55 UT
66mm APO refractor   127mm MCT x200
Nigel Longshaw   Peter Grego
(Chadderton, Oldham, UK) (Rednal, Birmingham, UK)

Independent visual observations of Venus by Nigel Longshaw and Peter Grego on 2007 May 2 and 7, recorded in integrated light. Many of the recorded features tally, proving that skilled visual observation still has an important role to play in Venusian studies

Some observers find features in Venus' upper atmosphere easier to discern others, not through a more acute sense of vision but because they have a greater retinal sensitivity to ultraviolet light, a wavelength in which Venus' cloud features are most easily seen (for more information, see equipment chapter, UV sensitivity and Venus' cloud features). To improve the visibility of Venusian cloud features, blue (Wratten 38A) and violet (Wratten 47) filters are helpful, and so too is yellow (Wratten 12 or 15). Often the planet has a faintly mottled appearance that is difficult to depict accurately on an observational drawing.

As with any other branch of observational astronomy, the observer needs to carefully attend to the object in view and get used to its appearance, and perhaps make a careful observational drawing on each occasion, rather than expect to see a variety of obvious features during the most cursory of glances through the telescope eyepiece. Venusian cloud features are best observed when the planet is a gibbous phase early in an eastern elongation or late on in a western elongation.

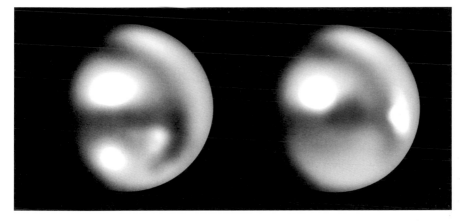

Apparent cloud changes on Venus observed by the author on the evening of 3 February 1996. The observations, made after sunset and separated by two hours, show subtle changes in the shape and intensity of the main cloud features. Peter Grego.

Venus' most prominent dusky features are usually seen near the terminator, from which they extend and fade, sometimes curving towards the poles. A distinct Y-shaped pattern of clouds is sometimes seen spanning the planet's equatorial region from the terminator towards the limb. Dusky collars bordering brighter polar regions are often seen, giving the impression of a planet with bright ice caps like Mars.

Venus' phase may be determined by several means. The most accurate way is to use an eyepiece micrometer with a graduated scale to measure both the planet's diameter and the widest part of the illuminated portion, and then divide the latter by the former figure to derive the phase. Alternatively, the planet's observed phase can be compared with a set of pre-drawn graphics showing various phases with known values. Least accurate of all, the drawing made at the eyepiece may be measured or the phase estimated with the eye alone.

(a)

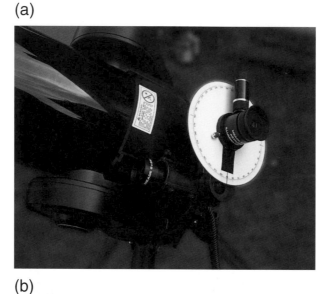

(b)

An illuminated reticle eyepiece such as this on the author's home-made micrometer enables the phase and phase angle of Venus to be accurately determined. Peter Grego.

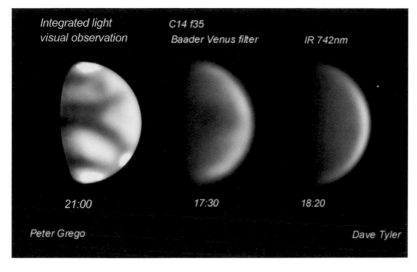

Comparison between a visual observation made in integrated light by Peter Grego and CCD images of Venus made through UV and IR filters by Dave Tyler on 5 May 2007.

Venusian Anomalies

The Schroeter Effect

Venus' predicted date of dichotomy (half-phase) does not always coincide with actual observations. When Venus lies to the east of the Sun dichotomy is sometimes observed to occur some days earlier than the predicted date; at western elongations, observed dichotomy sometimes occurs later than predicted. This phase anomaly is known as the 'Schroeter effect' after the lunar and planetary observer Johann Schroeter who noted it in the early 19th Century. It is caused by the scattering of sunlight along Venus' terminator; the effects of scattering are more pronounced nearer the planet's edge, where our view is directed through a thicker layer of Venus' atmosphere.

The Ashen Light

Venus' dark side has occasionally been observed to display a faint illumination when the planet is at a crescent phase and observed during astronomical night time. Known as the ashen light, the illumination is sometimes patchy and mottled rather than being an homogenous glow. There's little doubt that it is real, rather than an optical illusion, since it has been observed using an occulting bar within the eyepiece to remove the sunlit part of Venus from view. This curious phenomenon has proven difficult to explain, but recent theories as to its cause include the actual glow of Venus' hot surface, to lightning flashes within the Venusian atmosphere. Another phenomenon, this time one likely to be illusory, is the anti-ashen light, where during daylight observations the unilluminated hemisphere of Venus appears darker than the background sky.

Observing
Venus

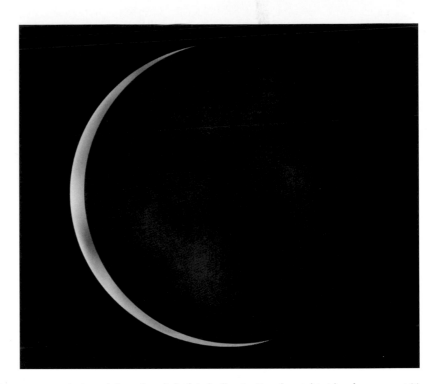

An exaggerated view of the ashen light faintly illuminating the night side of a crescent Venus.
Peter Grego.

Contour Anomalies

Images often show irregularities along Venus' terminator where dusky areas have
been underexposed. Since the eye has a far greater dynamic range than photographic
emulsion or even a CCD chip, the dense nature of the shading along the terminator
and its gradation, so apparent in many images, is frequently less apparent visually.
However, a combination of dark and/or bright features at the terminator can produce
a distinct unevenness in the terminator's curvature, producing projections along the
terminator or darkening or blunting of the cusps. Sometimes the north and south cusp
appear to have different curvatures – at dichotomy, the southern cusp is sometimes
blunter than the northern cusp. Anomalous bright cusp extensions have also been
observed from time to time, in addition to projections along the limb itself – both of
which may be attributable to Venusian atmospheric activity.

Recording Venus

Drawings of Venus are made on a 50 mm diameter circular blank. It is often difficult
to avoid producing a somewhat exaggerated impression of the tonal variation within
the planet's cloud features, since the features are usually very subtle and at the limit
of visibility. Intensity estimates of the cloud features can be recorded on a simple
line drawing to accompany the shaded observational sketch. The intensity estimates
scale runs from 0 to 5, 0 being extremely brilliant features, 2 the general tone of
the disk and 5 unusually dark shadings.

Transits of Venus

At inferior conjunction Venus moves between the Sun and the Earth; when the three bodies are precisely in line Venus can be observed to transit the Sun's disk as a perfectly black circular silhouette. Transits of Venus take place every 8, 121.5, 8 and 105.5 years. Only seven of them have taken place since the telescope was invented in the early 17th Century, the first having been observed telescopically from England way back in 1639. The last transit of Venus took place on 8 June 2004, and it proved to be one of the most widely observed astronomical events in history, viewable in its entirety from the UK, Europe, India and most of Asia. From eastern North America and much of South America, Venus was already on the Sun's disc as the Sun rose, while the transit was still in progress at sunset for observers in Japan and Australia. There is one more opportunity to view a transit of Venus during the 21st Century, and it takes place on 6 June 2012. The next pair of transits will occur in 2117 and 2125.

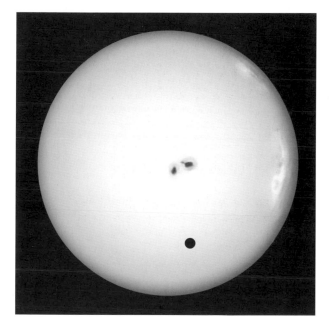

The 2004 transit of Venus, observed by the author through a 200 mm SCT and full aperture glass solar filter. Peter Grego.

Protected Unaided Eye Views

Although Venus covers just one-thousandth of the Sun's apparent area, it is so large – around one minute of arc across – that it can be seen by those with keen eyesight, provided that the eyes are safely shielded from direct sunlight by using a proper solar filter.

Regular sunshades, welder's glasses and sandwiched layers of photograph negatives are totally unsafe and should never be used. Special solar eclipse eye shades – sometimes called eclipse 'glasses' despite having no glass and no magnifying properties – have a thick layer of aluminised Mylar to prevent most of the

Sun's light, heat and ultraviolet radiation from reaching the eyes by reflecting it away. These shades can be used for brief but safe views of the Sun. Under no circumstances should these shades be worn as a filter to observe the Sun through the telescope eyepiece – the intense magnified energy of the Sun will quickly burn through them and cause permanent eye damage, if not blindness. Filters that fit over the eyepiece of a telescope must never to be used – they will heat up rapidly and shatter, with disastrous consequences. There are only two safe ways to observe the Sun through a telescope – careful eyepiece projection of the solar disk onto a shielded white card, and whole aperture filtration with a visually safe aluminised Mylar or glass filter.

The View in H-alpha

In H-alpha (hydrogen alpha) frequencies, the Sun's hot chromosphere is visible. Since the chromosphere lies above the photosphere (the body of the Sun visible in normal white light) observers viewing the transit in H-alpha light may be able to discern the Venusian disk a considerable time prior to first contact being seen in white light. Approaching the Sun, Venus may be seen silhouetted against the diffuse H-alpha glow (or more distinct features such as prominences) some time before it makes contact with the Sun's chromospheric limb. A period of several minutes separates first contact at the chromospheric and photospheric limbs (and also at last contact).

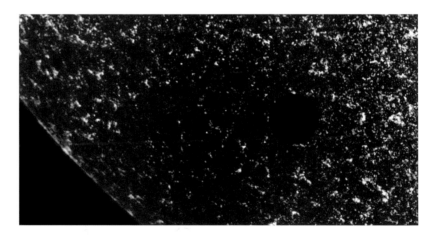

NASA's TRACE satellite imaged Venus' 2004 transit in H-alpha light. NASA.

The Venus Transit of 6 June 2012

The next transit will take place under almost opposite circumstances to that of 2004. The entire event, from first to last contact, will be observable from northwestern North America, the western Pacific, northeastern Asia, Japan, eastern Australia and New Zealand. High on most observers' dream list, if not wish list, is Hawaii, from which exotic location the Sun is almost directly overhead when the transit begins, and ends before the Sun sets over the Pacific. From most of North America the Sun sets while the transit is still in progress. Fortunately the last stages of the transit will be visible as the Sun rises above central Asia, most of western Europe (including the UK), eastern Africa and eastern Australia. Sadly, Portugal and much of Spain, western Africa and most of South America will be denied a view of the transit.

First contact (when the leading edge of Venus touches the edge of the Sun) occurs at around 22:09 UT (Universal Time). Observing the moment of first contact is by no means easy; the observer needs to pay attention to the point on the Sun's edge where Venus is due to make its entrance. In terms of position angle (measured in degrees starting from the northern point of the Sun's disk, eastward around the limb) Venus' first contact will be located at 40.7°. This figure needs to be translated into the orientation of the Sun as viewed through the telescope eyepiece or on the Sun's projected image. Through an astronomical telescope with an adequate full aperture solar filter, north is at the bottom of the image and east is on the right. With a diagonal eyepiece the image is flipped vertically, giving north is at top while east remains at right. On an image projected from the eyepiece of an astronomical telescope, north is at the bottom, east on the left hand side. Since the orientation of the Sun slowly swings from east to west during the day, its northern point rotating with respect to true north, the orientation of the Sun will vary depending on the time of observation and the observer's geographical location.

As Venus edges onto the Sun's disc, an unusual phenomenon can be observed. Refraction within Venus' dense atmosphere causes sunlight to be bent around the planet's following limb, producing a narrow arc of light extending around the edge of Venus. This tiny bow of light is fascinating to see – indeed, for many observers it represents the most memorable visual highlight of the transit.

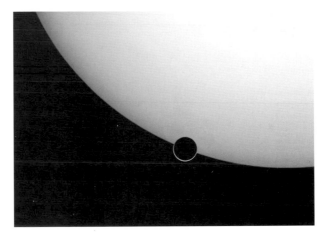

Author's observation of mid-ingress during the transit of 2004, showing the refractive arc around Venus' following limb. Peter Grego.

Because Venus is moving at an angular velocity of around four arcminutes per hour, it fully enters onto the Sun within 18 minutes of first contact; the point of second contact is the precise moment that the following limb of Venus last touches the edge of the Sun on ingress. A phenomenon known as the 'black drop' has often been reported in previous transits, producing some uncertainty in individual timings of the moment of second contact. The phenomenon gives the visual impression of a lingering ligament of darkness extending from Venus' following limb to the Sun's edge, making Venus appear like a drop of black ink suspended from the edge of the Sun – hence the name 'black drop'. The phenomenon is not an optical effect rather than a real one, caused by either poor seeing conditions, poor telescope optics, inaccurate focus or poor observing skills, or a combination of these factors. The best high resolution images of Venus at second contact during the 2004 transit show a perfectly clean second contact, with no hint of a black drop effect.

Venus reaches the other side of the Sun, after traversing a solar chord some 25 arcminutes long, a little more than six hours after it entered the solar disk. Because the orientation of the Sun changes gradually as it traverses the sky, the apparent track of Venus across the Sun will appear to be a curved line, rather than a straight one. Third contact occurs when the planet's preceding limb touches the Sun's western edge. As the planet leaves the Sun, another arc of refracted sunlight can be traced around Venus' preceding edge, caused in the same manner as that seen after first contact. Fourth contact happens at the very moment when the planet exits the Sun's photospheric disk, when the last traces of the Venusian silhouette disappears and the transit ends for viewers in white light. H-alpha viewers will be able to follow Venus for several minutes afterwards, through fourth H-alpha contact and possibly beyond.

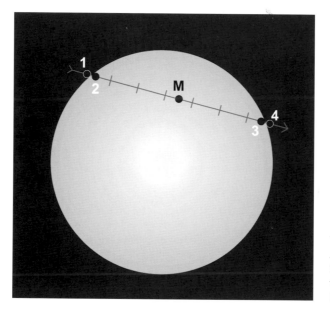

The transit of Venus, 6 June 2012. 1. First contact. 2. Second contact. M. Mid-transit. 3. Third contact. 4. Fourth contact.

Details of the transit of 6 June 2012

Contact	Time (UT)	Position angle
1st contact	22:09:29	40.7°
2nd contact	22:27:26	38.2°
Mid-transit	01:29:28	
3rd contact	04:31:31	292.7°
4th contact	04:49:27	290.1°

These details use geocentric data. The actual times will vary slightly depending on the observer's geographic location, but this amounts to no more than a couple of minutes' difference (either earlier or later) at most.

Stellar Occultations

Occultations of bright stars by Venus are extremely rare events. The following list involves bright 21st Century stellar and planetary occultations involving Venus:

Date	Time	Event
2035 Feb 17	15:19	Venus occults Pi Sagittarii
2044 Oct 1	22:00	Venus occults Regulus
2046 Feb 23	19:24	Venus occults Rho-1 Sagittarii
2065 Nov 22	12:45	Venus occults Jupiter

Societies, Groups, Useful Internet Resources and Bibliography of Worthwhile Books

Societies

The Society for Popular Astronomy (SPA)

Website: http://www.popastro.com

Address: The Secretary, 36 Fairway, Keyworth, Nottingham, NG12 5DU, United Kingdom.

Email: membership@popastro.com

Founded in 1953, the SPA is the largest astronomical society in the UK. It is aimed at amateur astronomers of all levels. Publications include the quarterly magazine, Popular Astronomy, and six News Circulars per year. The SPA hosts quarterly London meetings. The SPA has a thriving Planetary Section directed by Michael Hezzlewood.

The British Astronomical Association (BAA)

Website: http://www.britastro.org

Address: The Assistant Secretary, The British Astronomical Association, Burlington House, Piccadilly, London, W1J 0DU, United Kingdom.

A UK based astronomical association aimed at amateurs with an advanced level of knowledge and expertise. Its Mercury and Venus Section is directed by Dr Richard McKim.

The Royal Astronomical Society (RAS)

Website: http://www.ras.org.uk

Address: Royal Astronomical Society, Burlington House, Piccadilly, London, W1J 0BQ, United Kingdom.

Founded in 1820, the RAS is the UK's leading professional body for astronomy & astrophysics, geophysics, solar and solar-terrestrial physics, and planetary sciences. Its bimonthly magazine, Astronomy and Geophysics, features occasional informative articles about the planets. Fellowship of the RAS is open to non-professionals.

Association of Lunar and Planetary Observers (ALPO)

Website http://www.lpl.arizona.edu/alpo

This large association, based in the United States, has both a Mercury Section and a Venus Section, with numerous active planetary observing programs.

Unione Astrofili Italiani (UAI)

Website: http://www.uai.it/sez_lun/english.htm

Based in Italy, the UAI has active planetary observing sections, with a very informative English version of its website.

Internet Resources

Mercury Calculator

http://www.geoastro.de/skymap/MercuryVenus/mercury.html

Venus Calculator

http://www.geoastro.de/skymap/MercuryVenus/venus.html

USGS – Detailed Maps of Venus

http://planetarynames.wr.usgs.gov/vgrid.html

NASA's Planetary Photojournal

http://photojournal.jpl.nasa.gov/targetFamily/Mercury
http://photojournal.jpl.nasa.gov/targetFamily/Venus

Views of the Solar System

http://ftp.uniovi.es/solar/eng/homepage.htm

Books

Solar System Observer's Guide

By Peter Grego
Publisher: Collins
ISBN-10: 0540088277
ISBN-13: 978-0540088270
An observational guide, contains sections on observing Mercury and Venus.

The Compact NASA Atlas of the Solar System

By Ronald Greeley and Raymond M. Batson
  Publisher: Cambridge
  ISBN-10: 052180633X
  ISBN-13: 978-0521806336
  A detailed reference work containing charts of Mercury and Venus.

Russian Planetary Exploration: History, Development, Legacy and Prospects

By Brian Harvey
Publisher: Springer-Praxis
How Russia helped advance our understanding of the conditions on the Venusian surface.

Volcanic Worlds: Exploring the Solar System's Volcanoes

By Rosaly M.C. Lopes and Tracey K.P. Gregg
Publisher: Springer-Praxis
Includes explanations of the volcanic processes on Venus.

Exploring Mercury: The Iron Planet

By Robert G. Strom and Anne L. Sprague
Publisher: Springer-Praxis
A guide to the history and workings of the innermost planet.

Mercury Feature Index

Venus Feature index

Venus Feature Index

Subject index

Subject index

Printed in China